GW00373862

HOW MANY MOLEHILLS IN A MOUNTAIN?

HOW MANY MOLEHILLS IN A MOUNTAIN?

Measuring what you don't know in terms of what you do

MARCUS WEEKS

Ivy Press

First published in the UK in 2010 by
Ivy Press
210 High Street, Lewes
East Sussex BN7 2NS
United Kingdom
www.ivypress.co.uk

British Library Cataloguing-in-Publication Data
A catalogue record for this book is available from the British Library

ISBN: 978-1-907332-26-5

This book was conceived, designed, and produced by
Ivy Press

Creative Director **Peter Bridgewater**
Publisher **Jason Hook**
Art Director **Wayne Blades**
Senior Editor **Polita Anderson**
Designer **Glyn Bridgewater**
Picture Research **Katie Greenwood**

Cover image: Getty Images/Richard Lewis

Printed and bound in China
Colour Origination by Ivy Press Reprographics

10 9 8 7 6 5 4 3 2 1

While the author has taken all reasonable steps to ensure the accuracy of the information in this book, many measurements used have been rounded up or down for ease of calculation and for purposes of comparison. Similarly, average measurements are often only rough estimates or typical values, and sometimes wild generalisations. This book is intended as a guide to making rough measurements by comparison—it is not suitable for readers wanting an authoritative reference book.

CONTENTS

INTRODUCTION

'It don't mean a thing if it ain't got that ...' recognisable equivalent.

Ever wondered how big a blue whale really is? I mean, we all know it's big, but most of us never get to see one up close. So just how big are blue whales? Big as a house? Big as a plane? Bigger? You see, we instinctively look for something to compare them with to get an idea of scale – something familiar that we can relate to. Something like an elephant, for instance – they're pretty big, and almost everyone knows roughly what size they are. So if we're told that a blue whale's tongue weighs about the same as an African elephant, we know blue whales are not just big, they're BIG. About the same as 25 elephants, in fact.

That's the kind of thing this book is about – getting an idea of the size of things by comparing, rather than measuring, and bringing the big numbers down to a human scale. This means finding equivalents to things you don't know (generally things that are really huge or really tiny) in terms of what you do: everyday objects, household items, domestic animals or famous places and landmarks we can associate with. Instead of using units such as kilos (kg) or pounds (lb), litres (l) or pints (pt), or metres or (m) feet (ft), we replace them with units that make some sense to us, such as the weight and height of an average adult, the length of a football pitch and so on. From there, we can get a handle on some of the things we're a bit hazy about, like the weight of an elephant, and move on to the stuff we have difficulty getting our heads around. Not everyone has seen the Eiffel Tower (although we all know it's pretty tall) or visited Wyoming in the USA, but, put into context, by comparing them with the height of an average man or the size of a football pitch, they can be useful points of reference.

This is necessarily a somewhat inexact business, using average sizes and rounding numbers up or down to make calculations simpler, but it does help to give a rough idea of comparative sizes. And that's the key to understanding some of the very big (and very small) numbers that we're bombarded with every day.

It's difficult for us to imagine numbers greater than about a hundred. For example, what's a million? A million dollars? Wow, sounds a lot – but how much? Well, think of it this way: one dollar bill has an area of about 103 square centimetres (103cm^2) or 16 square inches (16in^2), so a million of them laid out flat would cover an area of around 10,322 square meters (10,322m^2) or 111,111 square feet (111,111ft^2). Huh? How about if we say that they would cover nearly two American football fields? Now that really makes sense to the sports enthusiasts. And for the rest of us, that million would take up the space of approximately 823 parked cars. That's a pretty big car park – and it works out at about 1,215 dollars a space.

We can be a bit hazy about everyday amounts, too. A sign says that a lift can manage a payload of one ton. So? How heavy is a ton? If we think of it as 12 guys and a 10-year-old boy we get a better idea.

Using things we know as units helps us to understand stuff we read in the newspapers, too. Statistics are meaningless (and dull) using standard units. Alarming reports of the depletion of the ozone layer or destruction of the rainforests don't mean a thing in square kilometres (km^2) or miles (mi^2), but when compared with the area of Wales we get the picture. Likewise, reports of pandemics leave us clueless, unless we can see the numbers in terms of something we know about – if 8.5 million people have gone down with the flu, we know that's not good, but if we're told that's like the population of New York City being laid low, we can see just how bad.

Another way to get things into an understandable perspective is to scale them up or down – this is useful when trying to work out very small or large concepts. For instance, we can get a better idea of just how small the nucleus of a hydrogen atom is if it is scaled up; even magnified 100,000 million times, it would be only about 2 millimetres (2mm) or seven-hundredths of an inch (0.07in) in diameter. On the same scale, a flea of about 3.8 millimetres (3.8mm) or fifteen-hundredths of an inch (0.15in) in life size would be 380,000 kilometres (380,000km) or 236,000 miles (236,000mi) long – almost big enough to stretch from Earth to the moon. Similarly, we can scale huge things, such as astronomical distances and sizes, down to manageable proportions. If Earth was the size of an orange, roughly 7cm (2¾in) in diameter, then the moon would be about 210cm (82½in) away – and the sun over 90km (56mi) away, centre-to-centre mean distances.

So, as you go through this book, marvelling at the comparisons between exotic and familiar objects, and record-breaking and everyday measurements, you'll also pick up units of height, length, weight, area etc. that provide some useful equivalents. And when you're faced with a new statistic – the distance of a satellite above Earth, say – you'll be asking questions such as 'How many Niles is that?' or 'What's that in Eiffels?' You might even start using your own units that have a more personal relevance.

Well, I do. I use the Camptown racetrack, which is 8km (5mi) long, or Pennsylvania Avenue—1mi 1,050ft from the White House to the Capitol. Come to think of it, that's about 2km; rather than a kilometre, though, I think of Regent Street, after the famous London shopping thoroughfare. I know my ideal weight is 20 tomcats, give or take a brick or two, and to keep in shape, I walk a Camptown, or swim a Regent a couple of times a week. I'm also trying to cut down my carbon footprint by using the car less, because I learnt that last year I drove it the equivalent of a North Pole to South Pole. At that rate I could do a lunar distance in just 10 years!

LENGTH & DISTANCE

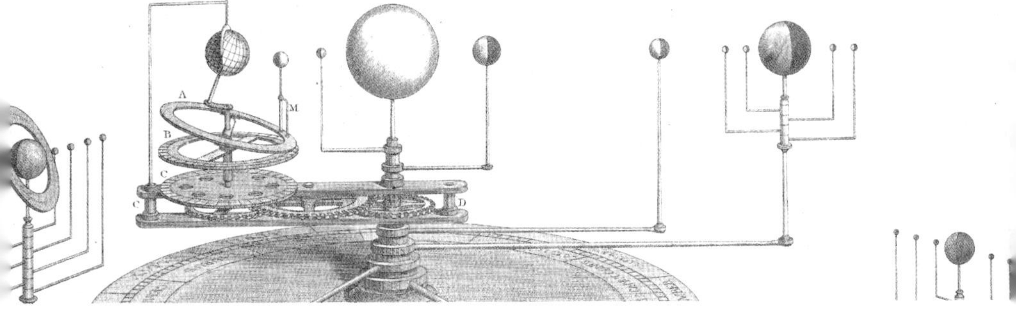

How long is a piece of string? We come across measurements of length or distance almost every day – not only in technical descriptions or in the news, but also in our everyday conversations. We need to know how far away some place is, or how long a table is or how big an animal is. If the answer comes back in kilometres or miles, metres or yards, chances are we'll only have the vaguest idea of what that actually means, and will have to refer to a map or a ruler to visualise the measurement. However, if we're told 'it's as long as …' or 'the size of a …', we get it straight away. This section looks at ways we can talk about lengths, from microscopic to astronomical, in units that make some sense to us.

RULE OF THUMB

The handiest things we have for comparison of short lengths are, well... hands. And feet, arms and legs, and length of a pace – in fact, many of our units of length started out as comparisons with parts of the body.

The foot is an obvious example of a comparative unit, but people have also measured in hands, spans (the width of an outstretched hand), cubits (the length of a forearm), digits (the width of a finger) and so on. The inch was originally the distance between the top of a man's thumb and the first knuckle – hence the "rule of thumb" meaning a rough measurement.

This is what we are interested in: rough measurement. Just about everyone has paced the length of a room to get an idea of its size in metres or yards. Standardisation of measurements (whether using a centimetre or inch for measuring), has led to increased accuracy of measurement, but this has meant that we've lost touch with some of the old rough-and-ready ways of saying how long something is.

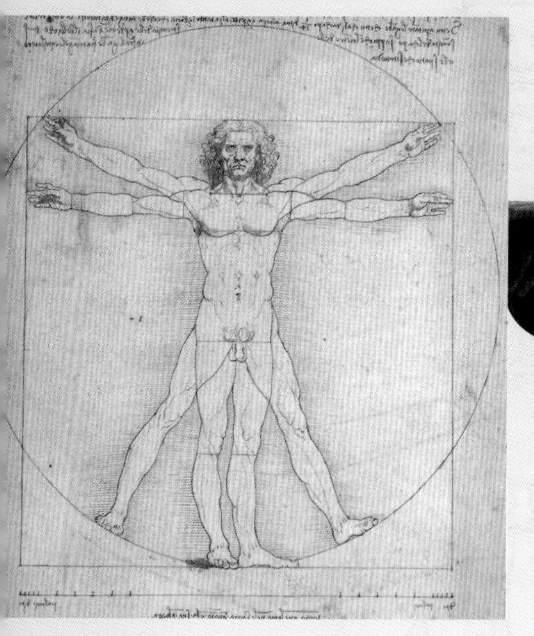

BLUE WHALE = APPROX 27.5M (90FT) – 15LEN

Ask an angler the size of the one that got away, and he won't tell you in centimetres or inches; he'll spread his arms, like that *Proportions of Man* drawing by Leonardo da Vinci. Now that's a unit we can all understand – let's call it the Leonardo (Len for short). The Leonardo comes into its own when describing even bigger fish such as the great white shark, and helps us put into perspective the size of creatures we may never come across such as dinosaurs and whales.

Little monsters: At the other end of the scale, even the monsters of the insect world are better measured with something smaller than the Leonardo. The span of a man's outstretched hand, for instance, is the distance between the tip of thumb and the little finger and measures around 23cm (9in). The wingspan of the white witch moth (the largest of the lepidopterans), can reach 1.25 spans, much the same width as the biggest spider, the Goliath birdeater, as measured by leg span.

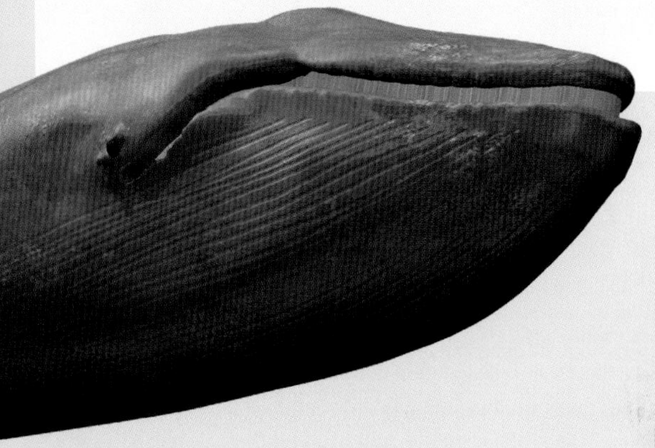

DIPLODOCUS = APPROX. 25.6M (84FT) – 14LEN

Winging it

Coincidentally, the Leonardo is just about the same as the wingspan of the bald eagle (and maybe American patriotic readers would prefer the B.E. to the Len as a unit of length). Although that's quite a span, it's not as big as the condor's 1.5 Leonardos, and is beaten hands down by the mighty albatross, which can stretch its wings up to 2 Leonardos. However, the all-time prize goes to the Quetzalcoatlus pterosaur. Fossil remains suggest that these creatures flew around North America in the Late Cretaceous period (see pages 86–87 to find out how long ago the period of the dinosaurs was) on a pair of wings a mighty 6 Leonardos across.

GREAT WHITE SHARK = APPROX. 11M (36FT) – 6LEN

A GIANT LEAP FOR MANKIND

In athletics it's called the long jump, but when compared to the leaps achieved by other animals, it's really not that long. The farthest anybody has jumped is 8.95m (29ft 4½in), achieved by Mike Powell at the World Athletics Championships in Tokyo in 1991. The distance is about 9 paces, or 4.5 Leonardos, which sounds pretty impressive – it's three-quarters the distance achieved by that legendary leaper the kangaroo (6 Leonardos). True, but not so impressive measured against two other famous jumpers ...

Although frogs can only manage up to about 4 paces,

and fleas a miserable 1.5 spans, if we think of their body length and scale that up to human size, their leaps make ours look pretty puny. At about 100 times its body length, the frog's best jump is the equivalent of a man leaping 200 paces. The flea's mighty bound, 150 times its body length, is equivalent to a 300-pace Olympic record. That's difficult to Imagine ... but easier if we think in terms of something else: the length of an average parking space – about 5 paces – which we can call the "parked car" (or Pcar for short). So, an Olympic jumper with a flea's leaping ability would be able to clear a line of 60Pcars.

Cheater

The prize for giant leaps goes to a cheater (no, not a cheetah; he's in Section 8: Speed). The flying fish doesn't actually 'fly', but leaps out of the water and then glides on specially developed fins. This 'fin-assisted' leap can be as long as 400 paces, or 80Pcars. Now *that's* impressive.

In the long run

Another impressive athletic feat in ancient Greece was Pheidippides's run from the battlefield at Marathon across the mountains to announce the Greek victory in Athens, a distance of about 40km (25mi). The Greek hero died on the spot, but he had just run 240km (150mi) over the previous two days. Marathon races inspired by his final trip are a bit longer – officially a distance of 42.195km (26mi 1,155ft).

Even using the parked car as a unit, that's not easy to picture. Sure, it's a long way to run, but just how far is it?

Now, it often happens when you ask directions that your destination is a '5-minute walk' away. While that's almost invariably an underestimation in practise, it provides a useful unit of distance. Assuming a moderately brisk walking speed, the '5-minute walk' (or F.M.W.) is about 500 paces, or 100Pcars.

So, to answer the question, a marathon is approximately 85F.M.W. – in other words it would take about 7 hours to walk.

Compare the marathon with another legendary crossing, poet Lord Byron's swim across the Hellespont (now known as the Dardanelles) in Turkey. This channel, which separates Europe and Asia, sounds more of a challenge than it actually is, being only 2.5F.M.W. at its narrowest. It's chicken feed to modern swimmers, who prefer the challenge offered by the Straits of Gibraltar (28.5F.M.W.) or the Straits of Dover in the English Channel (68F.M.W.) – though most people wanting to get to France opt, these days, for the Channel Tunnel (see page 15).

GAME, SET & MATCH

Sports events offer a rich source of lengths and distances that can be used for comparison. As well as the obvious 100m sprint and the 400m track, there are all kinds of sport fields, grounds and pitches that have standard lengths. The football pitch, for instance, at 105m (344.5ft) or the American football field of 110m (360ft), depending on your preference; or a tennis court at 23.77m (78ft), or even a basketball court at 28–29m (92–94ft).

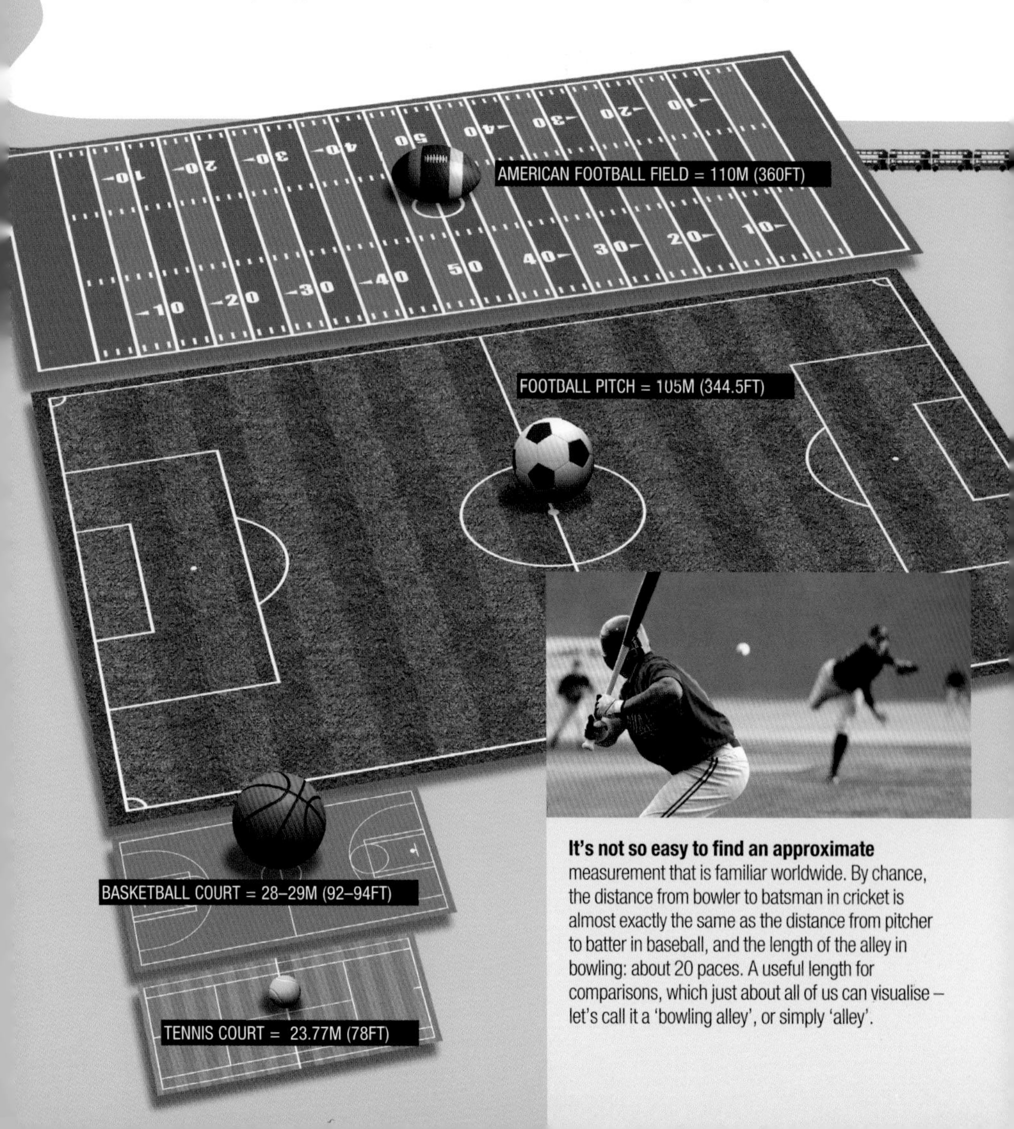

It's not so easy to find an approximate measurement that is familiar worldwide. By chance, the distance from bowler to batsman in cricket is almost exactly the same as the distance from pitcher to batter in baseball, and the length of the alley in bowling: about 20 paces. A useful length for comparisons, which just about all of us can visualise – let's call it a 'bowling alley', or simply 'alley'.

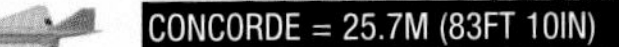

KNOCK NEVIS SUPERTANKER = 458M (1,502FT)

HUGHES FLYING BOAT = 66.65M (218FT)

Way to go

A peculiarly English unit of size for comparisons, the London bus, gained an almost international usage. The red double-decker was as familiar through films and TV shows as that other London icon, Big Ben. Alas, the Routemaster (the double-decker most people think of) has almost disappeared, but lives on as a measure of 30ft (so two would fit neatly on a bowling alley). The American equivalent, the 45ft Greyhound bus, has still to replace it.

Other forms of transport vary widely in length, from little boats to supertankers, lorries to trains, and biplanes to jumbos. The longest of them all (to date) is the eight-engined, 682-car freight train, which travelled from Port Hedland, Australia, in June 2001: it was 7.353km (4.6mi) long, or about a 1¼-hour walk from end to end.

BROOKLYN BRIDGE = 1,818M/1.13MI (5,989FT) – 365PCARS

Over & under

City dwellers who have been caught in rush-hour traffic know just how long a line of parked cars is, and if they work in London, New York, San Francisco, or Sydney, will have a shrewd idea of how long their most famous bridges are. For the rest of us, the picture here may help us sympathise. Luckily for commuters between France and England, the Channel Tunnel is a railway tunnel.

GO WITH THE FLOW

The marathon provides a handy length to measure longer distances, and is particularly useful for describing that beloved topic of general knowledge quizzes, the length of rivers. The two main contenders for longest river, the Nile and the Amazon (the Nile wins, by the way, but only by about 10 marathons), are 150 marathons or more long, as is the Mississippi–Missouri system. Not far behind are the Yellow River (130 marathons), the Mekong (116 marathons) and the Congo–Chambeshi (112 marathons) – all of which are longer than the width of the United States.

Going places

Strangely, it might help us visualise these long distances better if we look at them in terms of something even bigger – and that means looking at some global distances. For example, even looking at a map of the Nile gives us no real idea of how long it is, but if we know that it's the same distance as New York to Prague, and that the flight between those cities is just over 8 hours, it begins to make sense. These international distances become even more meaningful when seen on the globe. When you can see that the radius of Earth is just about the same length as the Nile, things take on a new perspective.

- **Nile** 6,650km (4,132mi) = 160 marathons
- **Amazon** 6,270km (3,900mi) = 150 marathons
- **Mississippi–Missouri** 6,275km (3,902mi) = 150 marathons
- **Huang** 5,464km (3,398mi) = 130 marathons
- **Congo–Chambeshi** 4,700km (2,922mi) = 112 marathons
- **Volga** 3,692km (2,294mi) = 88 marathons
- **Murray–Darling** 3,370km (2,094mi) = 80 marathons
- **Rhine** 1,320km (820mi) = 32 marathons
- **Seine** 776km (482mi) = 18 marathons
- **Thames** 346km (215mi) = 8 marathons

Statistics

- 8,851.8km (5,500mi 1,755ft) total length
- 6,259.6km (3,889mi 2,640ft) of sections of actual wall
- 359.7km (223mi 2,640ft) of trenches
- 2,232.5km (1,387mi 1,320ft) of natural defensive barriers such as hills and rivers

The extent of the longest man-made object in the world, the Great Wall of China, takes some imagining, even if you look at a map or visit it. In total, including all its branches, it's a staggering 8,851.8km (5,500mi 1,755ft) long. To give you a better idea, that's about 1.3 Niles – more than twice the width of Australia or the USA, and even longer than the continents of Africa or South America – and if straightened out, it would take nearly 11½ hours to fly from end to end.

Some of its length (about 1.7 Rhines) consists of natural defences such as hills and rivers, and there is about a Thames of trenches, but that still leaves almost exactly an Amazon of real wall.

GREAT WALL ON USA

GREAT WALL ON AUSTRALIA

GREAT WALL ON AFRICA

GREAT WALL ON S. AMERICA

GOING TO GREAT LENGTHS

To get some idea of the distance between planets in the solar system, imagine the sun is in San Francisco, and the Kuiper belt, containing dwarf planets such as Pluto and Haumea at the edge of the solar system, runs through New York. On that scale, Earth would be at Sacramento (just 130km/80mi away), passing Mercury and Venus on the way. Our nearest neighbour, Mars, would still be in California, and the next, Jupiter, would be in the centre of Nevada. Approximately 160km (100mi) east of Salt Lake City you would find Saturn, and Uranus lies in deepest Nebraska. Continue on through Indiana to Ohio, you would find Neptune halfway between Indianapolis and Columbus, still a good way from our final destination, the Kuiper Belt, New York.

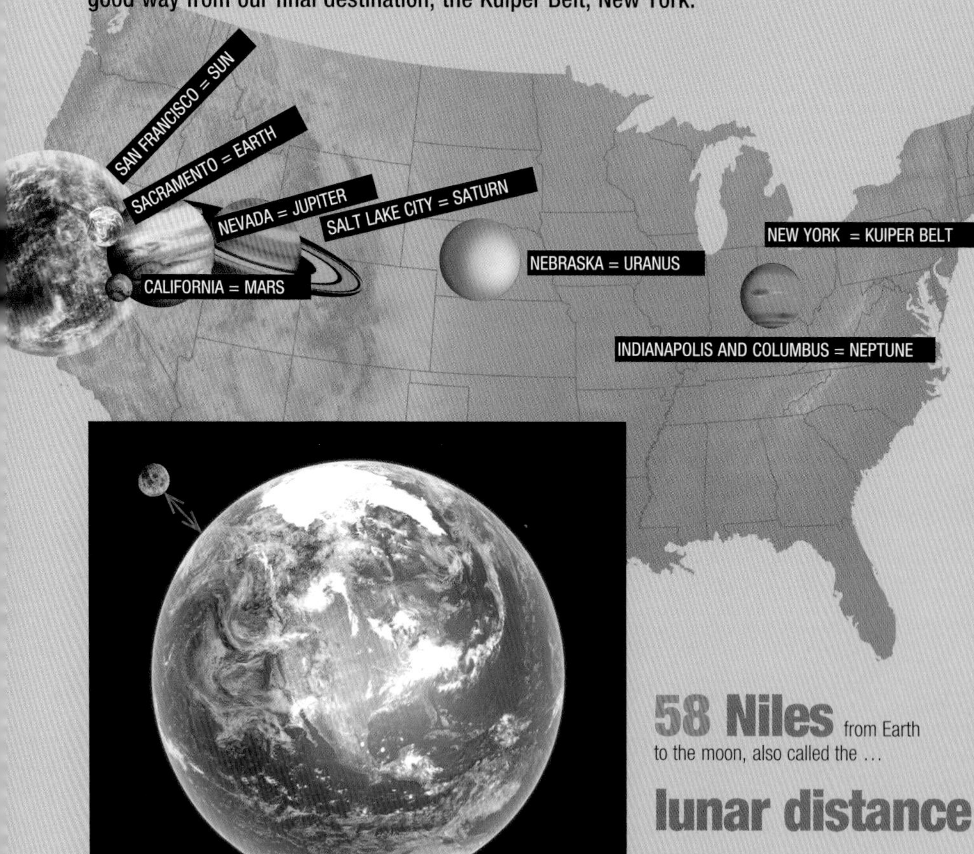

58 Niles from Earth to the moon, also called the …

lunar distance

A giant leap for mankind, but a short step when compared with the vast distances outside our cosy little world.

Having a ball

To get an idea just how big the solar system is, try this little experiment.

Get hold of a bowling ball with a diametre of about 25cm (8in). This will represent the sun. Now, how far away do you reckon Mercury will be at this scale? The answer may surprise you – it's 10 paces. The next planet, Venus, is another 9 paces, putting us almost a bowling alley away from our bowling ball. Next is Earth, 7 paces later (and incidentally it would be about the size of a peppercorn in this model), followed by Mars after another 14 paces. Here's where the surprises really start: It's 95 paces to the next planet, Jupiter (the biggest in our solar system; about the size of the top of a man's thumb, on this scale). Keep walking – there's a long way to go yet. There's 112 more paces to reach Saturn, then 249 paces from there to Uranus; another 281 paces takes us to Neptune, and you finally reach the dwarf planets of the Kuiper belt (so small a pinhead would be too big for them), in another 242 paces.

In all, you'll have covered over 1,000 paces from the sun to the edge of the solar system. Using the same scale, our nearest neighbouring star, Proxima Centauri, would be a really long walk – about 6,900km (4,300mi) away.

How high is the moon?

In the game of finding and naming comprehensible units, astronomers are way ahead of us. Okay, they don't talk in terms of 'How many Niles to the moon?' (about 58, since you asked), but they do use the average distance from Earth to the moon, known as the lunar distance. Similarly, the average distance to the sun gets a fancy name – the astronomical unit, or AU (equivalent to about 16,833 Great Walls) – and can be used to measure some of the vast distances in space.

For the really big distances, they use the light year, the distance light travels in one year. That's about 63,240AU – and from that you can calculate that it takes the light from the sun 8 minutes 20 seconds to reach Earth.

Our solar system, at about 70–80AU across, may seem vast, but is lost in the immensity of our galaxy, the Milky Way, which is something like 100,000 light years from side to side. And the universe? Astronomers reckon it's at least 93 billion light years across. And counting.

70–80 AU

is the size of the solar system

100,000 light years

is the size of our galaxy (Milky Way) from side to side

93 billion light years

is the size of the universe (increasing all the time)

A HAIR'S BREADTH

When someone has had a narrow escape, it's often described as a "hair's breadth' from disaster, but what exactly is a hair's breadth: the thickness of human hair varies enormously at about 0.04–0.25mm (0.0016–0.0098in), so is a far from accurate unit. For measuring very small distances, we need something a bit more consistent, and familiar – so, how about using the thickness of this page?

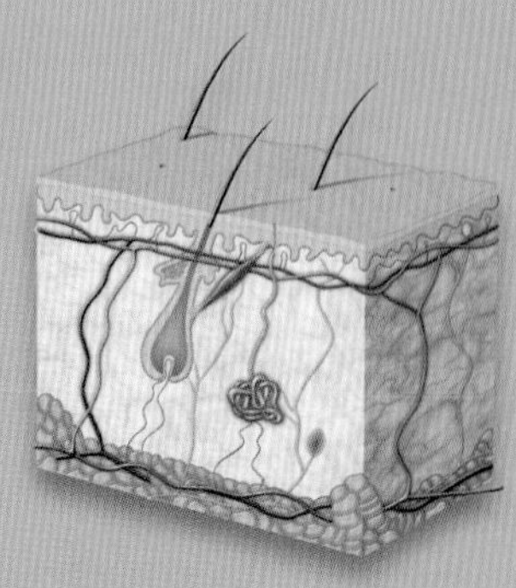

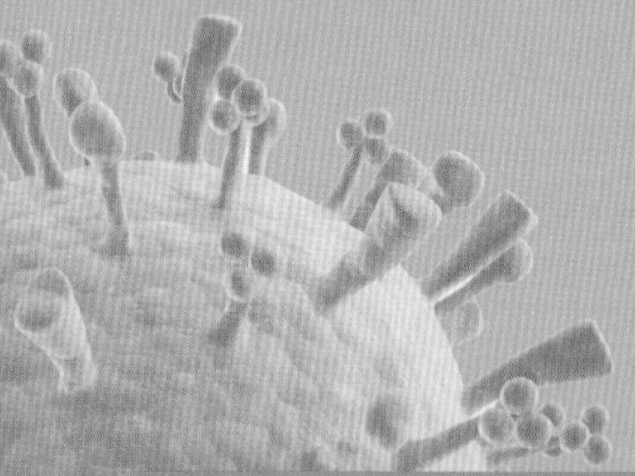

Using a page's thickness as our yardstick, we can see just how small some things really are. Amoebas, for instance, range in size from about 800 to 15 micrometres – which doesn't tell us a lot, until we know that even the biggest are only half of the page's thickness, and about a thousand of the smallest would fit side by side in the same space. If you think that's small, consider viruses, which are about 1,000 times smaller than the amoebas (between 10 and 300 nanometres to be more precise) – and we haven't even begun to get down to things at a molecular or atomic level . . .

The core of the matter

Compared with viruses, the byword for smallness, the flea, is a positive giant at up to a 20-page thickness in length (hold all the pages from the beginning of this book to page 40 together to see what that looks like). Even their eggs are about 3 pages. Let's take the flea as a unit – a medium-sized flea of around a 13-page thickness.

Compare that with something truly microscopic:

Hydrogen nucleus x 100,000,000,000 = flea

In other words, the nucleus of a hydrogen atom would have to be magnified 100,000,000,000 times to be the size of a flea. Magnified the same amount, the hydrogen atom as a whole would be about 20 parked cars across – and our 'hair's breadth' would vary in thickness from about 4,000km (2,500mi) to 25,000km (15,500mi). On the same scale, that flea would be 380,000km (236,000mi) long – almost big enough to stretch from Earth to the moon.

2 AREA

Trying to work out the wide open spaces: Visualising lengths and distances is hard enough, but when it comes to area, most of us are in the dark. Farmers might be quite happy talking about hectares (h) or acres (ac), and estate agents about square metres (m^2) or square feet (ft^2), but we need some kind of comparison to find out what on earth they mean. Whether it's a question of working out how much turf you need for that lawn, or worrying just how big the hole in the ozone layer is, it helps to have something to measure it against.

$$A = \iint_D \sqrt{\left(\frac{\partial f}{\partial x}\right)^2 + \left(\frac{\partial f}{\partial y}\right)^2 + 1}\, dx\, dy.$$

BACK TO SQUARE ONE

It isn't easy finding that dream house. The photos from the estate agent are bound to be misleading, so you have to look at the dimensions – and if they're in square metres or square feet, you're probably none the wiser.

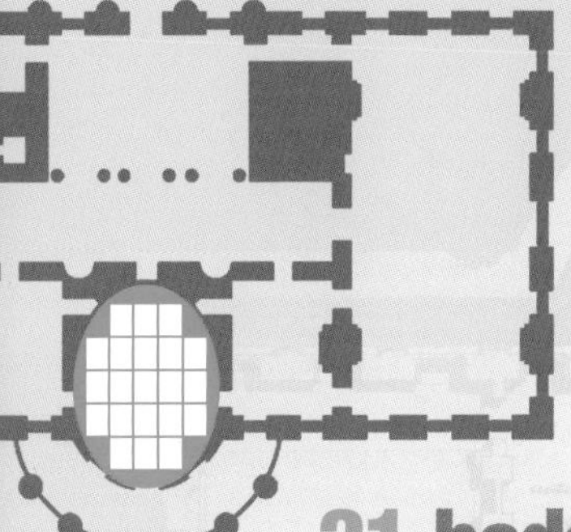

What you need is something to help you visualise those areas. For example, it says the bedroom is 15m^2 (about 160ft^2): what would that look like with a double bed in it? Now, a standard double bed is about 2.5m^2 (27ft^2), so you could fit six of those in this bedroom – or one, with plenty of room to walk around.

The garden is a good size, too: 42m^2 (450ft^2). That's enough room for about 17 double beds – think of it as an area of 4 x 4 beds, plus a bit extra for the barbecue.

21 beds in 57m^2 (614ft^2)

A sporting chance

Just as the bowling alley provides us with a convenient unit of length (see page 14), sports facilities can help us think about areas. Possibly the most useful candidate is the boxing ring (the red square in the diagram, below right): easy to visualise and get an idea of scale from the boxers in it. It's a useful size, too: about 16 double beds. Using the boxing ring (B.R. for short), we can get a handle on some other familiar areas, such as a football pitch (170B.R.s), or an American football field (125B.R.s).

Sport's not your thing?

Think parked cars. The normal parking space is around 10m^2 (108ft^2), roughly one-quarter of a boxing ring – so the parking space (P-space for short) is another handy unit to work with. So when you see a figure of, say, 1,250m^2 (13,500ft^2), you know that's 100P-spaces, or the equivalent of a car park for 100 cars.

Or the area of an Olympic-sized swimming pool, if you prefer.

ALBERT HALL = 5,400M^2 (58,000FT^2) – 540P-SPACES – 135B.Rs

COLOSSEUM, ROME = 23,225M^2 (250,000FT^2) – 2,322P-SPACES – 580B.R.s

GREAT PYRAMID = 53,095M^2 (571,530FT^2) – 5,309P-SPACES – 1,327B.R.s

PENTAGON = 117,355M^2 (1,263,240FT^2) – 11,735P-SPACES – 2,933B.R.s

Out in the open

Using units such as the boxing ring (B.R.) – the red square in the diagram below left – and the parking space (P-space), we can start making sense of some bigger ground areas: places most of us only read about such as the Great Pyramid of Giza (5,309P-space, or 1,327B.R.s), or the Colosseum in Rome (2,322P-space, or 580B.R.s). And really big buildings, too, such as the Pentagon – equivalent to a 11,735-space car park. Although John Lennon may not have known how many holes it takes to fill the Albert Hall, we know it takes 135 boxing rings.

TRAFALGAR SQUARE 12,100M^2 (130,200FT^2) – 1,210P-SPACE – 1.64 FOOTBALL PITCHES

PIAZZA SAN MARCO 12,128M^2 (130,500FT^2), TRAPEZOIDAL DIMENSIONS: 175M (574FT) LONG, 56.6M (185FT) AND 82M (270FT) WIDE – 1,213P-SPACE – 1.65 FOOTBALL PITCHES

RED SQUARE 23,100M^2 (248,600FT^2) – 2,310P-SPACE – 3 FOOTBALL PITCHES

TIANANMEN SQUARE 440,000M^2 (4,736,000FT^2) – 4,400P-SPACE – 60 FOOTBALL PITCHES

We can also use the same measurements for public spaces such as Red Square in Moscow (2,310P-space/3 football pitches) or St Mark's Square in Venice (1,213P-space/1.65 football pitches). That helps us get something like Central Park into perspective: never mind that it's 341 hectares (1 hectare = 1,000P-space), or 843 acres (1 acre = 400P-space) – it's 637 American football fields, which is nearly eight times bigger than Tiananmen Square.

WHAT A STATE

It gets a bit problematic to find comparisons when it comes to geographical areas. The unit favoured by British journalists is an area the size of Wales, but that means nothing to non-Brits. Americans might want to use one of their states – Wyoming, for example, because it's square(ish) and conveniently covers an area of very roughly 250,000km^2 or 100,000mi^2. That means that there are about 40 Wyomings in a United States. Other candidates include Colorado (only very slightly bigger than Wyoming), Texas (2.75 Wyomings) and Alaska (almost 7 Wyomings).

However, we're in danger of continuing the rift of units used on opposite sides of the Atlantic – metric vs US standard measurements, and even USA vs UK. So instead of specifying too rigidly, let's use local units wherever possible: there are islands all round the world that fit the bill.

A world of difference

The total surface area of the world is 510,072,000km^2 (196,939,000mi^2). That's a heck of a lot of Wyomings or Madagascars – it's even 50 times the total of the USA. To make any sense, we need to break that huge figure down: 70 per cent of it is water, so the remaining land area is only about 150,000,000km^2 (57,500,000mi^2). Still baffled? Well, think of it as seven continents: Asia, Africa, North America, South America, Antarctica, Europe and Australia. Or 15 Canadas, or Chinas, or USAs – they're all the same size, give or take a bit. Or 47 Indias, 70 Saudi Arabias or 250 Iberian Peninsulas …

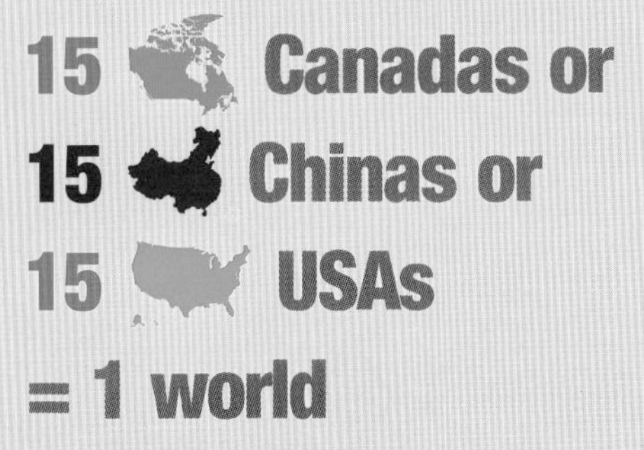

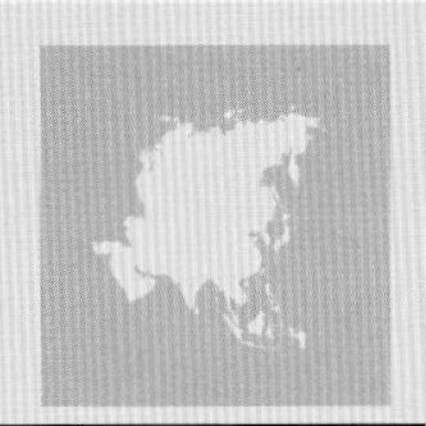

ASIA 44,579,000KM^2 (17,212,000MI^2)

AFRICA 30,221,530KM^2 (11,668,598MI^2)

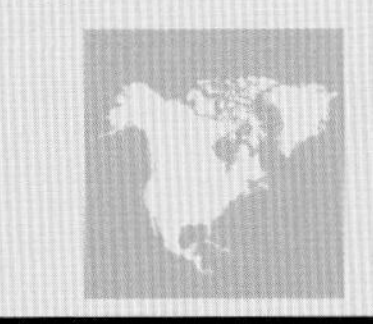

N. AMERICA 24,709,000KM^2 (9,540,000MI^2)

S. AMERICA 17,840,000KM^2 (6,890,000MI^2)

ANTARCTICA 14,000,000KM^2 (5,400,000MI^2)

EUROPE 10,180,000KM^2 (3,930,000MI^2)

AUSTRALIA 7,617,930KM^2 (2,941,299MI^2)

THE SEVEN SEAS

Seventy per cent of the earth's surface is water, and a visitor from outer space could be forgiven for thinking it was even more if he approached from over the Pacific Ocean. This huge expanse of water – 166,266,877km^2 (64,196,000mi^2) – is greater in area than all the world's land together. The Atlantic Ocean is only about half that, but still covers more than all its bordering continents, Africa, Europe, North America and South America.

The other three oceans are no puddles either: the Indian Ocean could swallow Africa and Asia, the Southern Ocean could cover Africa and South America with room to spare, and the Arctic Ocean is roughly the same area as Antarctica (isn't that pleasingly symmetrical!).

Then there are all the other smaller seas, ranging from the Arabian Sea (about 1.8 Saudi Arabias) through the Mediterranean Sea (3.9 Frances or 4.2 Iberias) to the Gulf of California (1.5 Cubas).

The Pacific Ocean seen from the International Space Station.

Oceans

1 **Pacific Ocean** 166,266,877km² (64,196,000mi²). *Greater than the earth's total land surface area – by about 20,000,000km² (7,722,043mi²)*
2 **Atlantic Ocean** 86,505,603km² (33,400,000mi²). *Could swallow up all bordering continents – Africa, Europe, North America and South America*
3 **Indian Ocean** 73,555,662km² (28,400,000mi²). *Equivalent to Africa and Asia*
4 **Southern Ocean** 52,646,688km² (20,327,000mi²). *A bit more than Africa and South America*
5 **Arctic Ocean** 13,208,939km² (5,100,000mi²). *About the same as Antarctica*

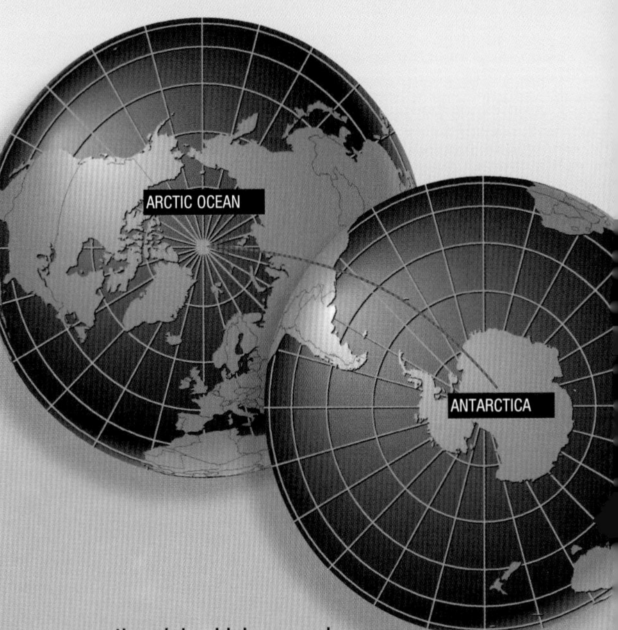

Water, water, everywhere …

As well as all the seas and oceans, there are countless inland lakes, ponds and pools. There's some dispute over which is the biggest. Although it's considered by many to be a lake, the Caspian Sea (the clue's in the name) is technically a small ocean, but at 1.5 Wyomings it's certainly the largest landlocked body of water. Another controversial contender, Michigan–Huron, comes second at 0.5 Wyomings, but only if you think it's a single body of water – otherwise it's Lake Superior (one-third of a Wyoming).

Caspian Sea = 1.5 Wyomings

Lakes

1 **Caspian Sea** 371,000km² (143,000mi²) = 1.5 Wyomings
2 **Michigan–Huron** 117,702km² (45,445mi²) = 0.5 Wyomings
3 **Superior** 82,414km² (31,820mi²) = 0.3333 Wyomings
4 **Victoria** 69,485km² (26,828mi²) = 3.4 Waleses, just over 1 Sri Lanka
5 **Tanganyika** 32,893km² (12,700mi²) = 1.6 Waleses, about 2 Sri Lankas
6 **Baikal** 31,500km² (12,200mi²) = 1.5 Waleses

JUST DESERTS?

In contrast to the seas and oceans, there are also vast areas of the world where there is no water at all. The Sahara, for example, which is more than 3.5 times the size of the Mediterranean Sea and covers about one-third of Africa. Just over the Red Sea is the Arabian Desert at just under 1 Med – about twice the size of the Gobi, or 3 Kalahari deserts.

However, those dry places pale in comparison with the frozen wastes of the polar ice caps, even though the latter are shrinking at an alarming rate. The Antarctic and Arctic ice sheets each cover about 1.5 Saharas. Or 55 Wyomings, if you must.

Can't see the wood for the trees

The polar ice caps aren't the only areas getting smaller: the world's rainforests are also being cut down to make way for more profitable business. Estimates of the depletion vary, from 2 American football fields to 600 parking spaces every second, or an area 30 times the size of the Pentagon every minute. In other words, we're losing approximately 3 Waleses (one-quarter of a Wyoming, for American readers) of rainforest every year – maybe even more.

Compare this with the Hundred Acre Wood frequented by Winnie the Pooh in children's stories. An equivalent size of rainforest is flattened every minute-and-a-half.

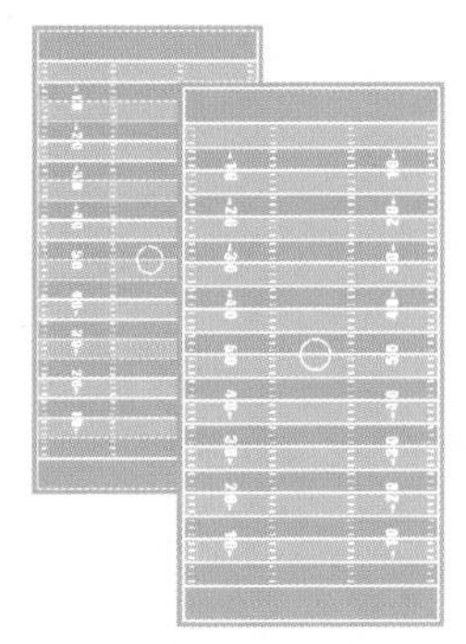

2 football fields per second

A hole in one

What about the ozone layer? We all know there's a hole in it – but how worryingly big is it? Well, at its largest, in September 2006, its average size was twice the area of Antarctica, and a fair bit bigger than North America. The good news is that it seems to be gradually shrinking now.

27,453,874km^2 (10,600,000mi^2)

From 21–30 September 2006, the average area of the ozone hole was the largest ever observed.

ON THE BUTTON

If you look around you, you'll find all kinds of things that can be used for comparisons of area – everyday items that we see every day. A shirt button, for example, is about 0.155in^2 (which just happens to be close enough to 1cm^2), so is useful for talking about quite small areas. Then there's a standard CD or DVD: about 12cm (4¾in) across, working out to about 113 shirt buttons. Or an average-sized dinner plate, roughly four times the area of a CD.

If you find it difficult dealing with circular areas, what about the page you're looking at? It's about 2.5 CDs, or 60 per cent the area of a dinner plate. Or you might prefer a sheet of A4 paper (about 1.25 dinner plates or 5.6 CDs). Laid out flat, 40 of them would cover a double bed, and 160 a parking space.

x 113 =

SHIRT BUTTON (11MM/0.43IN/ DIAMETER) = 1CM^2 (0.155IN^2)

CD/DVD (12CM/4.72IN DIAMETER) = 113.1CM^2 (17.5IN^2)XXX

x 4 =

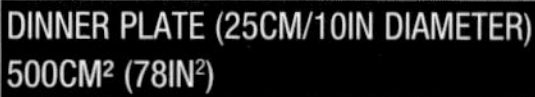

DINNER PLATE (25CM/10IN DIAMETER) 500CM^2 (78IN^2)

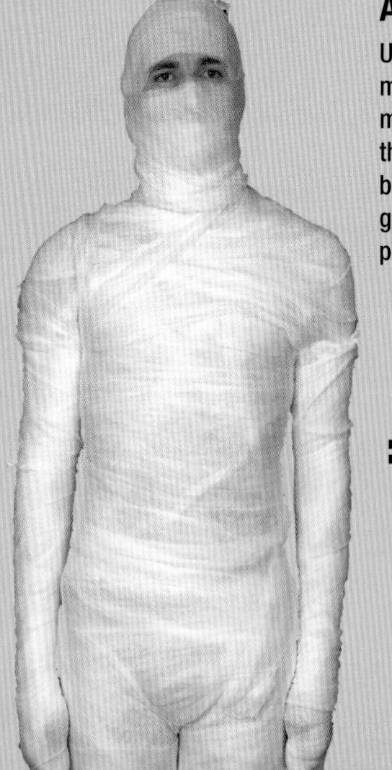

All wrapped up

Using everyday units gives us a better idea of apparently meaningless statistics. The surface area of an average adult male is about 1.75m^2 (18ft 9in^2) – so what? Knowing that this is the same size as 28 sheets of A4 paper (or 62 pages of this book) is more useful, especially if for some reason you've been given the task of wrapping a body in paper. Or it's 35 dinner plates, if you're planning a cannibalistic dinner party.

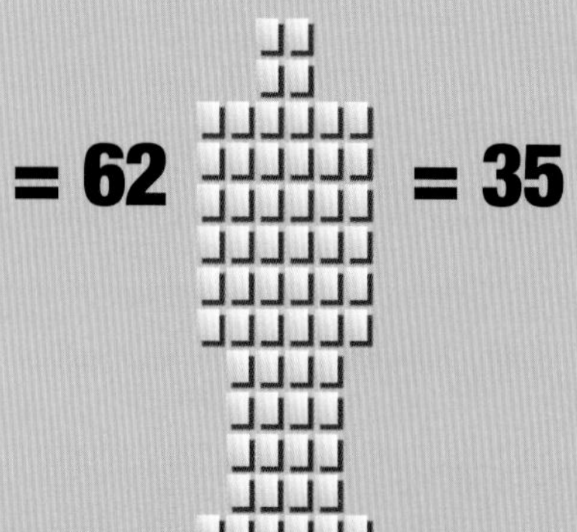

How many angels can dance on the head of a pin?

When it comes to microscopic areas, the head of a pin is a pretty good benchmark to start from – it's about one-hundredth of a shirt button. Or you could say it is six times the area of a full stop in Helvetica 12 point (.), which in turn is 50 times the cross section of an average human hair. Our pinhead is equivalent in area to 30,000 human red blood cells. If you think that's small, what about the HIV virus: it takes about 5,000 of them to cover the same area as a red blood cell.

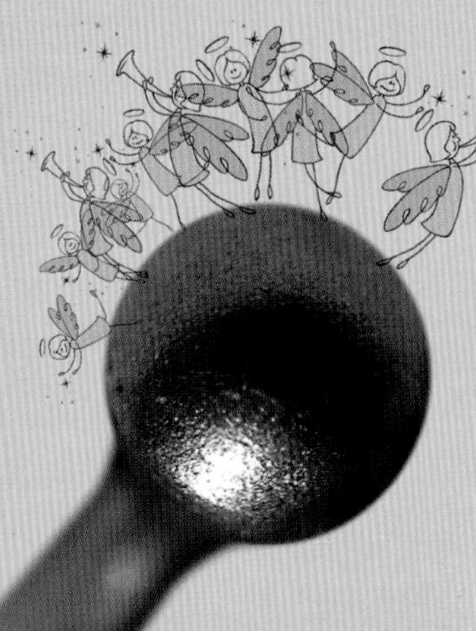

So … if we assume, just for argument's sake, you can fit 10 dancing angels on a pinhead, then they would each take up the area of 3,000 red blood cells, in case you were wondering.

= 30,000 

human red blood cells

Table of microscopic comparisons

- **1 shirt button** = 100 pinheads
- **1 pinhead** = 6 Helvetica full points in 12 point (or 10 angels)
- **1 12-point Helvetica full point** = 50 cross sections of average human hair
- **Cross section of 1 human hair** = 100 red blood cells
- **1 human red blood cell** = 5000 HIV virus particles

To infinity & beyond

Getting down to really small measurements highlights the usefulness of working with equivalents rather than standard units of measurement. It's okay talking about microscopic lengths in thousandths of an inch (although it's difficult to say, and even more difficult to type), but it gets really cumbersome when we're talking thousandths of a square inch or whatever. The figures have so many zeroes that you end up having to count them just to make sure. It's not much better using the International System of Units (SI) either, even though there are those cool prefixes such as micro-, nano-, pico- and (my favourite) zepto-: it can end up sounding like the cast list of a Marx Brothers film.

A STORMY SEA

The surface area of the moon (that's near side plus far side) is approximately 13½ times smaller than that of Earth, and is covered in craters – some as small as a pinhead, but the biggest about the size of Iceland. It's got seas, too, but these are really great waterless plains. The largest of these, the Oceanus Procellarum, or Ocean of Storms, is much the same size as the Arabian Sea: a bit over half the size of the USA, or 20 Wyomings.

A spot of trouble

There's a storm on Jupiter as well. An enormous anticyclonic storm that looks like a giant red spot just below Jupiter's equator, called (unimaginatively) the Great Red Spot of Jupiter. Its size fluctuates from around three-quarters the area of the moon to just a bit bigger than the total surface area of Earth, and it's kind of oval in shape these days, but it seems to be getting slowly smaller and more circular. One day, it may disappear completely. Wouldn't that be a shame?

3

HEIGHT & DEPTH

River deep, mountain high: Because we know our own height quite accurately, and maybe the heights of people we know well, we can estimate similar heights with a fair degree of precision. Where we need a little help is in translating that knowledge to things such as tall buildings or mountains, when we're told their height in metres or feet. It's even worse when it comes to depths. We know what the shallow and deep ends of our local swimming pool are, but the Grand Canyon? It's so much easier if you've got something familiar to compare these measurements with, some rough units of height and depth. Read on …

MEASURING UP

Most of us live in towns or cities, surrounded by buildings and structures of various sizes, so it's natural that we think of heights in those terms. Most towns and cities have one or two landmarks – usually the tallest in the area – which people get their bearings from. Depending on where you live, this could be one of the famous ones such as the Big Ben clock tower, Empire State Building, Eiffel Tower or CN Tower. Or it could be just another office block. Whatever it is – it won't mean much without some kind of standard to compare it with.

That's another story: The height of buildings is often described by the number of floors or levels. Although that varies from building to building, it does help when making comparisons. As a rough guide, a story is the equivalent of 1.5 M. Eiffels.

Upwards trend

Let's take a famous landmark, the Eiffel Tower in Paris, and see how it measures up. It's 324m (1,063ft) tall – about 300m (1,000ft) if we ignore the mast on top. To put that into human perspective, it's 177 times the height of its designer, Monsieur Eiffel (or simply MEiff for short), assuming he was a well-built 1.83m (6ft) tall.

Now compare that with some iconic buildings (the list changes as records are broken), or the one in your neighbourhood, and you've got a good basis for interpreting heights.

What an Eiffel

Eiffel Tower to top of TV mast: 324m (1,063ft) – 177MEiff – 1 Eiffel

Clock tower housing Big Ben: 96.3m (315ft 11in) – 52.5MEiff – about one-third of an Eiffel

Empire State Building, without mast: 381m (1,250ft) – 208MEiff –1.3 Eiffels

Taipei 101: 509.2m (1,671ft) – 278.5MEiff – 1.7 Eiffels

Warsaw, radio mast: 646.4m (2,121ft) – 353.5MEiff – 2.2 Eiffels

CN Tower, Toronto: 553.3m (1,815ft) – 302.5MEiff – 1.8 Eiffels

Burj Khalifa: 828m (2,717ft) – 453MEiff – 2.75 Eiffels

Sears Tower, Chicago, without mast: 443m (1,454ft) – 242.5MEiff – 1.37 Eiffels

Monumental heights

When a film maker wants to let his audience know the action's in London, he uses the cliché of red buses, Big Ben's clock tower or Nelson's column. Similarly, for New York he'd use the Empire State Building or Statue of Liberty, for Egypt the Great Pyramid in Giza, and for Rio de Janeiro the statue of Christ the Redeemer on the Corcovado Mountain.

However, you don't have to travel the world to find out just how tall they are. Using the Eiffel Tower and the Monsieur Eiffel (MEiff) for comparison, you can get a better idea, without leaving your armchair.

MT RUSHMORE HEADS, ABOUT 18M (60FT) – 10MEIFF

GREAT PYRAMID, GIZA, 146M (481FT) – ABOUT HALF AN EIFFEL – 80MEIFF

MOTHERLAND STATUE, VOLGOGRAD, 85M (279FT) – 46.5MEIFF

STATUE OF LIBERTY 46M (151FT) + PEDESTAL = 93M (305FT) – 50MEIFF (STATUE ITSELF IS ONLY 25MEIFF)

CHRIST THE REDEEMER, RIO DE JANEIRO, 39.6M (130FT) – 21.5MEIFF (INCLUDING 9.5M/31FT PEDESTAL)

NELSON'S COLUMN: STATUE OF NELSON 5.5M (18FT) (3MEIFF) STANDS ON TOP OF A 46M (151FT/25MEIFF) COLUMN = 51.5M (169FT) – 28MEIFF

MAKING WAVES

From a human perspective, buildings such as the Eiffel Tower and the Burj Khalifa seem pretty impressive – and they come in handy as units of height. However, they're not that spectacular when compared to heights in the natural world.

Take waves, for instance. When an avalanche at the foot of the Lituya Glacier in Alaska tumbled into the sea in 1958, it caused a wave 524m (1,720ft) high – that's about 1.75 Eiffels.

ANGEL FALLS

979m (3,212ft)

Waterfalls

The Angel Falls in Venezuela, the world's highest waterfall, is actually half an Eiffel taller than the Burj Khalifa, at 3.26 Eiffels. In fact, there are plenty of waterfalls around the world that can be measured in Eiffels. Sadly, Niagara is not one of them. At just 50.9m (167ft) high, the Eiffel Tower towers above the falls, and they are best thought of as around 29 M. Eiffels.

BURJ KHALIFA

m (2,715ft)

MT EVEREST = 8,848M (29,029FT) – 29.5EIFFEL – 10.7B.K.

CERRO ACONCAGUA = 6,962M (22,841FT) – 23.2EIFFEL – 8.4B.K.

MT MCKINLEY = 6,194M (20,320FT) – 20.6EIFFEL – 7.5B.K.

MT ELBRUS = 5,642M (18,510FT) – 18.8EIFFEL – 6.8B.K.

MONT BLANC = 4,808M (15,774FT) – 16EIFFEL – 5.8B.K.

KILIMANJARO = 5,895M (19,341FT) – 19.7EIFFEL – 7.2B.K.

A = Burj Khalifa
B = Eiffel Tower
C = Petronius Platform

A B C

MAUNA KEA = 10,203M (33,474FT) FROM SEABED – 34EIFFEL – 12.3B.K.

Making molehills out of mountains

The Eiffel and the Burj Khalifa come into their own when it comes to looking at mountains. For city dwellers, skylines such as those of New York and Paris contain the tallest things they're likely to see, but their Empire States and Eiffels are pretty puny when put alongside some of the world's peaks.

Mount Everest, in the Himalayas, is the highest mountain above sea level at 29.5 Eiffels, or 10.7 Burj Khalifas (B.K.), but it has a rival in Mauna Kea, Hawaii, most of which is below the surface of the Pacific Ocean. Although only 4,207m (13,803ft) or 14 Eiffels above sea level, its peak is a massive 34 Eiffels or 12.3 Burj Khalifas from the seabed – compare this with a similarly placed man-made construction, once claiming to be the tallest in the world, the Petronius Platform in the Gulf of Mexico: just over 2 Eiffels from its top to the seabed. Pathetic!

LITUYA GLACIER WAVE

524m (1,720ft)

EIFFEL TOWER

324m (1,063ft)

GREYHOUNDS & OTHER ANIMALS

Although the red double-decker London bus has almost disappeared, it is still used as a measure of height (and length, see page 15). At about 2.5 M. Eiffels (MEiff), it remains a handy comparison – better, perhaps, than the modern Greyhound bus at around 1.6 MEiff.

Using M. Eiffel and a bus as a guide, we can look at the heights of animals – we know they're tall, but we're not sure how tall. The giraffe, for example, can grow taller than a London bus, and at 3 MEiff is much the same height as a mature Tyrannosaurus rex.

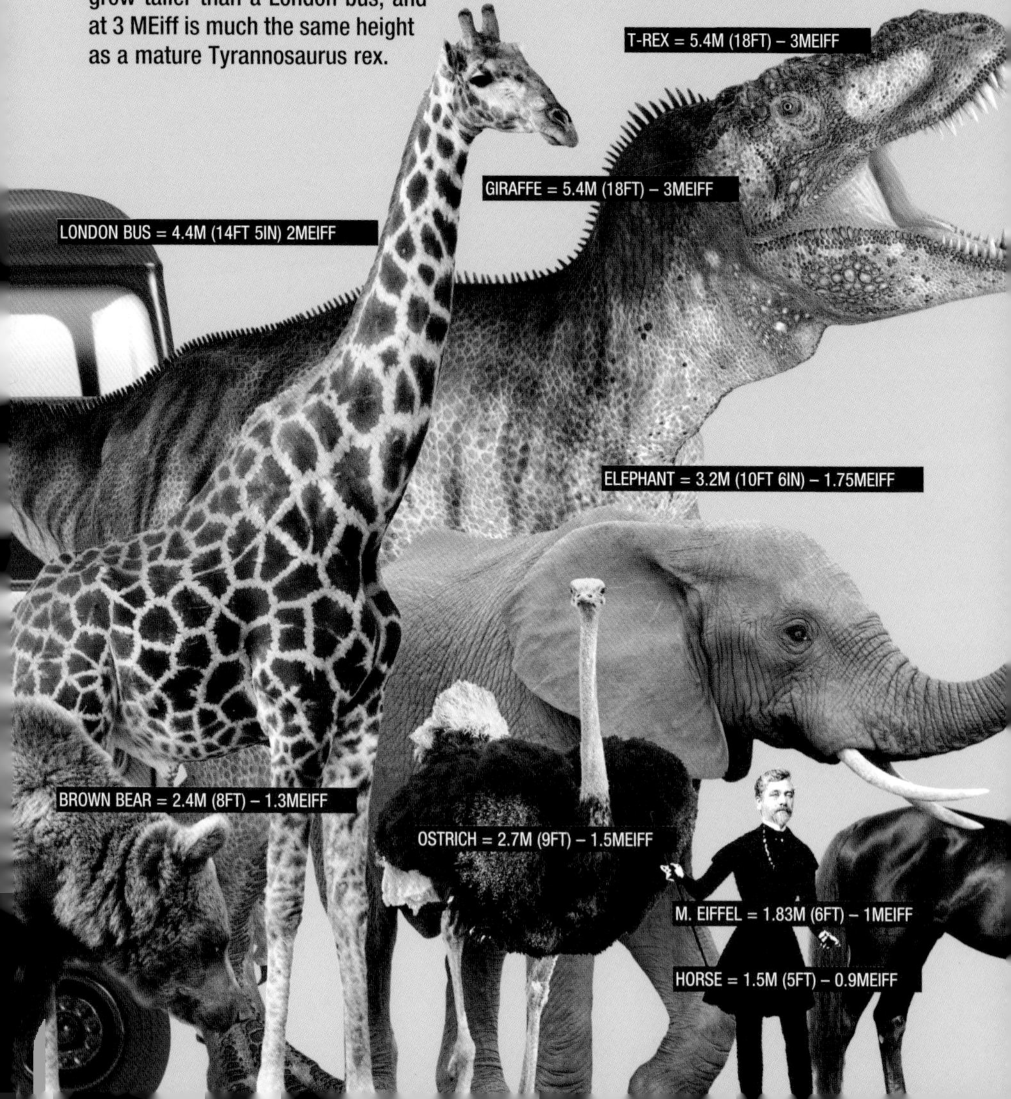

T-REX = 5.4M (18FT) – 3MEIFF

GIRAFFE = 5.4M (18FT) – 3MEIFF

LONDON BUS = 4.4M (14FT 5IN) 2MEIFF

ELEPHANT = 3.2M (10FT 6IN) – 1.75MEIFF

BROWN BEAR = 2.4M (8FT) – 1.3MEIFF

OSTRICH = 2.7M (9FT) – 1.5MEIFF

M. EIFFEL = 1.83M (6FT) – 1MEIFF

HORSE = 1.5M (5FT) – 0.9MEIFF

Bound to succeed:
The athlete on page 12 with the jumping prowess of a flea would do well not only in the long jump but also in the high jump. A flea can jump more than 100 times its body length into the air; this would be a leap of 100 MEiff, which would easily clear the Great Pyramid of Giza. A Tyrannosaurus rex with that ability could jump over the CN Tower in Toronto.

Seeing the wood for the trees

One of the problems with plants is that they keep on growing. Some trees have lived for hundreds, maybe thousands, of years and show no signs of stopping. So it's difficult to measure them accurately, but we can get a reasonable idea of a kind of everyday height for mature trees of most species. Other plants vary in size so much, depending on their age and environment, it's virtually impossible to pin down an average.

Of course, there are some record breakers such as the Californian redwood (up to 21 giraffes), some species of bamboo (up to 7MEiff), the Saguaro cactus (8.5MEiff) and the occasional 4-MEiff sunflower; or, at the other end of the scale, bonsai trees as small as 2.54cm (1in).

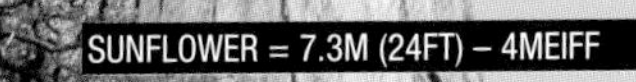

OAK = 40M (132FT) – 22MEIFF

SAGUARO CACTUS = 15.5M (51FT) – 8.5MEIFF

THE DEEP BLUE SEA

The seas of the world vary in depth about as much as mountains do in height. In fact, the deepest part of the oceans, the Mariana Trench in the Pacific Ocean, pretty much compares with the highest mountain: at its deepest, it's about 1.25 Everests. Put another way, if Everest had its base at the bottom of the Mariana Trench, its peak would be about 7 Eiffels below the surface.

However, that's an extreme. The average depth of the Pacific Ocean is a mere 15.5 Eiffels, and the Atlantic, Indian and Southern oceans are all around 13 Eiffels on average. The Arctic can only manage an average of 4 Eiffels, with 17 Eiffels at its deepest point.

Compared to the oceans, even the deepest lake hardly dents the surface. Averaging around 1.5 Eiffels, Lake Baikal in Siberia holds the record with a maximum depth of 5.5 Eiffels.

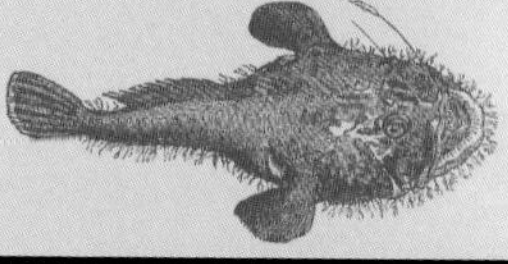

Hidden depths: At the bottom of the oceans, the pressure of water can be the equivalent of 50 or so jumbo jets, and the sun's light never penetrates that far, so you wouldn't expect to find much living down there. Surprisingly, there are creatures that apparently thrive at these depths. Even on the floor of the Mariana Trench, about 11.3km (7mi) down, there are tiny one-celled organisms called foraminifera, and giant tube worms are found nearly as deep.

Human diving achievements not so noteworthy: OK, we've managed to get a manned bathyscaphe to the very bottom, but only once; and with an atmospheric diving suit we've reached about 2.5 Eiffels, but the deepest scuba dives aren't even as much as 1 Eiffel.

Scratching the surface

All this talk of the highest and deepest points on the earth's surface can be somewhat misleading. Compared to the size of the earth, Everest is less than a pimple, and the Mariana Trench is just a small scratch.

THE DEPTHS OF THE EARTH

The Grand Canyon, carved out of rock by the Colorado River, is famously more than 1.6km (1mi) deep in places. Some claim it's the deepest canyon in the world; others say that's the Tsangpo Canyon (AKA Yarlung Zangbo Grand Canyon) in Tibet, or the Copper Canyon in Chihuahua, Mexico. Since they're still arguing, we can assume they're all about the same – about 1,000 M. Eiffels, or 5.7 Eiffel Towers (Eiffels).

5.7 Eiffels = 1 Grand Canyon

Dead in the water: Although it's a mere Eiffel deep, the Dead Sea (or Salt Sea – it doesn't matter, as it's technically a lake anyway) is notable for the fact that its shores are the lowest place on dry land in the world: 1.4 Eiffels below sea level.

Water eats away the rock under the ground, forming caves and tunnels. The deepest of these such as the cave systems in Abkhazia, Georgia, can reach 7 Eiffels from their entrances to their floors, deeper than the Grand Canyon cuts into the earth's surface.

Skin deep

None of the natural or man-made holes in the ground goes any more than skin deep, geologically speaking. As yet, we haven't managed to get below the earth's crust, which varies in thickness from about 0.6 to 8 Everests. And (sorry, Jules Verne) there's a long, long way to go before we get to the centre.

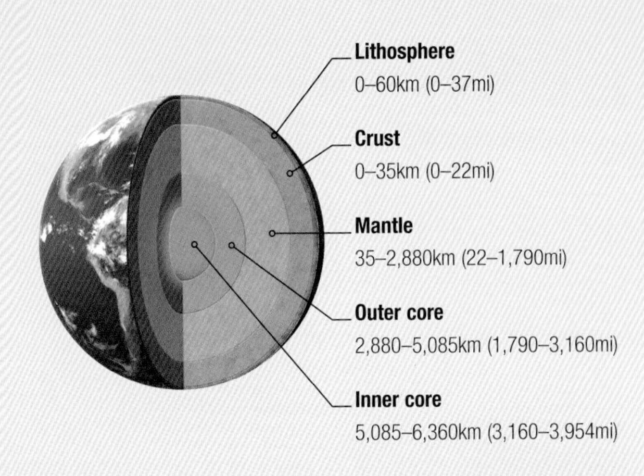

Digging deep

It seems that we just love digging. Not only in the garden, or laying pipes in the road, but on a large scale, too. The bigger the better. This primeval urge to dig has given us some remarkable achievements: the Channel Tunnel, for instance, which connects England and France and goes as deep as 17 London buses below sea level. However, it's not the deepest – oh no; that honour goes to the Eiksund Tunnel in Norway, nearly an Eiffel below sea level at its lowest point.

We go to extraordinary depths to extract mineral resources, too. Mines in South Africa can be as deep as 130 Eiffels, and there's an open pit mine in Bingham Canyon, Utah, that goes down a mighty 4 Eiffels. The record for drilling, however, goes to the aptly named Kola Superdeep Borehole in northern Russia. Scientists there bored 1.4 Everests down into the earth's crust in 1989. Why? Because we just love digging, I suppose.

FLYING HIGH

When you look up and see a jet plane in the sky, it looks really high up, which it is, of course – most passenger jets cruise over 9,144km (30,000ft) above the ground. At that height, they would clear the peak of Everest with an Eiffel Tower to spare. While that's the highest most people are likely to get, it doesn't seem quite so exciting if you think of it as distance on the ground: it's only about 9.14km (5.7mi). Not even one-quarter of a marathon.

Propeller-driven planes and hot-air balloons don't even make it that high, and would crash into Everest about halfway up. No, the high flyer is the gas-filled balloon. Manned gas balloons have reached up into the stratosphere nearly 4 times the height of Everest, and unmanned weather balloons nearly 6 times its height.

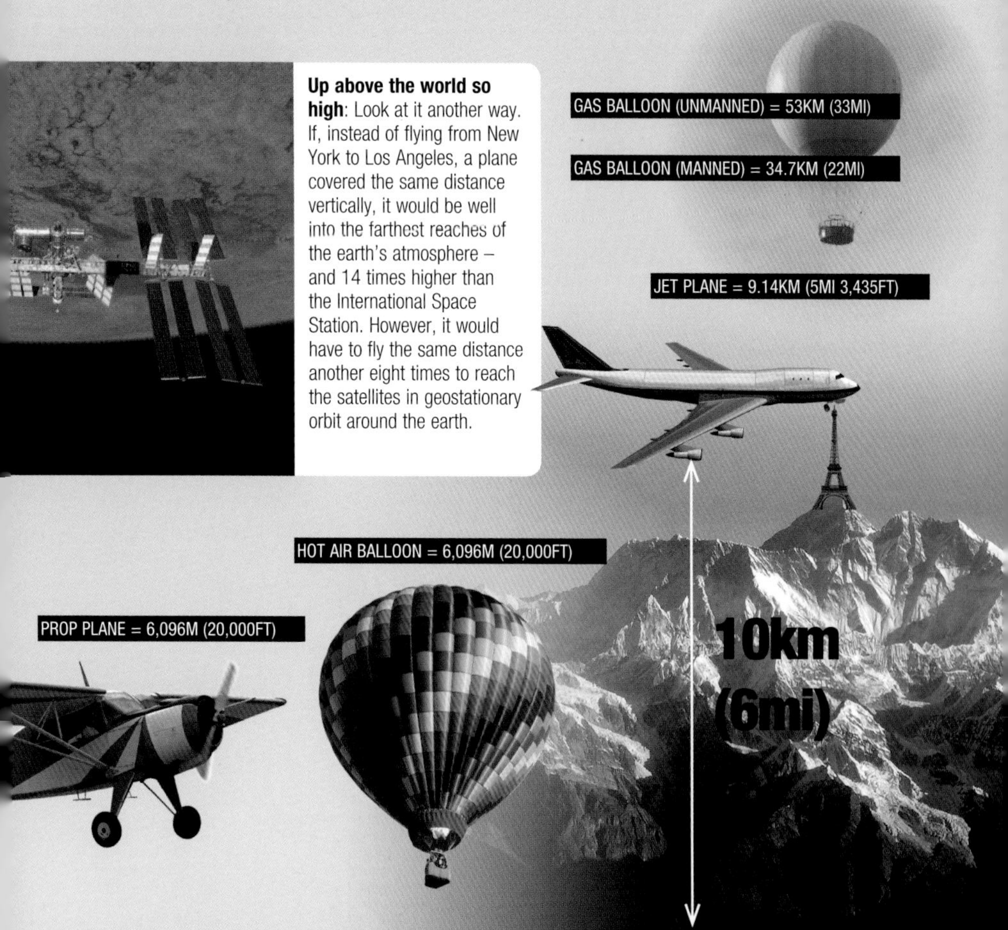

Up above the world so high: Look at it another way. If, instead of flying from New York to Los Angeles, a plane covered the same distance vertically, it would be well into the farthest reaches of the earth's atmosphere – and 14 times higher than the International Space Station. However, it would have to fly the same distance another eight times to reach the satellites in geostationary orbit around the earth.

4 WEIGHT, MASS & DENSITY

A weight off your mind: There are any number of everyday objects that we almost instinctively know the weight of. We can guess the weight of a cake because we're used to picking up sacks of sugar or flour of specific weights. We're also quite good judges of how heavy people are, too, simply because we have a good idea of our own weight. So, it's only a small step to gauging the weight of larger or smaller things. Which, in turn, can be used as units to measure other things we're not sure about. Simple. Well, comparatively simple.

THE AVERAGE MAN

A lot of our measurements of weight evolved from units that were practically useful: sacks of grain, bags of cement and so on. These days, few of us have to heave bags of cement about very often, so the idea of a hundredweight (110lb or roughly 50kg) isn't very useful for our purposes. What we have instead is our own body weight. Discounting extremes of obesity and starvation, we can imagine an average man's weight (around 80kg or 175lb) – let's call it a John Doe. That's a good starting point for making comparisons with other weights.

Kilogrammes and pounds: These can be confusing, so a unit for comparison of smaller weights would be helpful. We don't have to look too far to find something to fit the bill – an average-size tomcat would be about right. In fact, a John Doe weighs about 20 tomcats (which we can abbreviate to J.D. and Tom, or T.C. respectively).

Pretty soon, you'll be looking at weights in a new light, and asking, "What's that in tomcats?" or "How many John Does in this?"

How many elephants in a blue whale?

Using the John Doe (J.D.) and the tomcat to give an idea of scale, less familiar animals start to make more sense. Pygmy shrews, for example: they're tiny. So tiny that it would take 1,980 of them to balance the scales with a tomcat. At the other end of the scale, there are elephants. People often talk facetiously about something weighing as much as an elephant, but it's not such a silly idea. An adult African elephant weighs about 75 John Does, so it's a nice round figure to add to our scale of units. And from there we can size up the real monsters such as the blue whale, which weighs in at a mighty 25 elephants.

101 dalmatians: Dogs may make perfect pets and working animals, but they're useless for measuring with. The trouble is they come in all shapes and sizes, from chihuahua to great dane. Take dalmatians. An adult can be anything from 4 to 8 tomcats in weight. Nevertheless, if we take an average of 6 tomcats, we can calculate (roughly speaking, of course) what those 101 dalmatians weighed altogether: 606 tomcats, about 30 John Does or 3 cows – which is about the same as an adult hippopotamus.

25 elephants in a blue whale

1 mammoth = 2 elephants/28 horses

HOW MANY KOALAS IN A SUMO WRESTLER?

Using a couple of established units of weight – the John Doe (the weight of the average human adult male) and the tomcat (the weight of an average adult cat) – we can start working with derived units to get a perspective on some less common things. For instance, a professional sumo wrestler weighs in at around 160kg (350lb). Now, that doesn't mean much until you think of it as 2 John Does.

So far, so dull – but remember that 1 John Doe = 20 tomcats, so our sumo champ tips the scales at about 40 tomcats. Now, how does that relate to other animals – a koala, for instance? Well, once we know that a koala weighs much the same as 2.5 tomcats, we can apply a little bit of simple maths:

Weight of sumo wrestler in tomcats = 2 x 20 = 40

Divide this by 2.5, and we have the equivalent in koalas – 40 ÷ 2.5 = 16

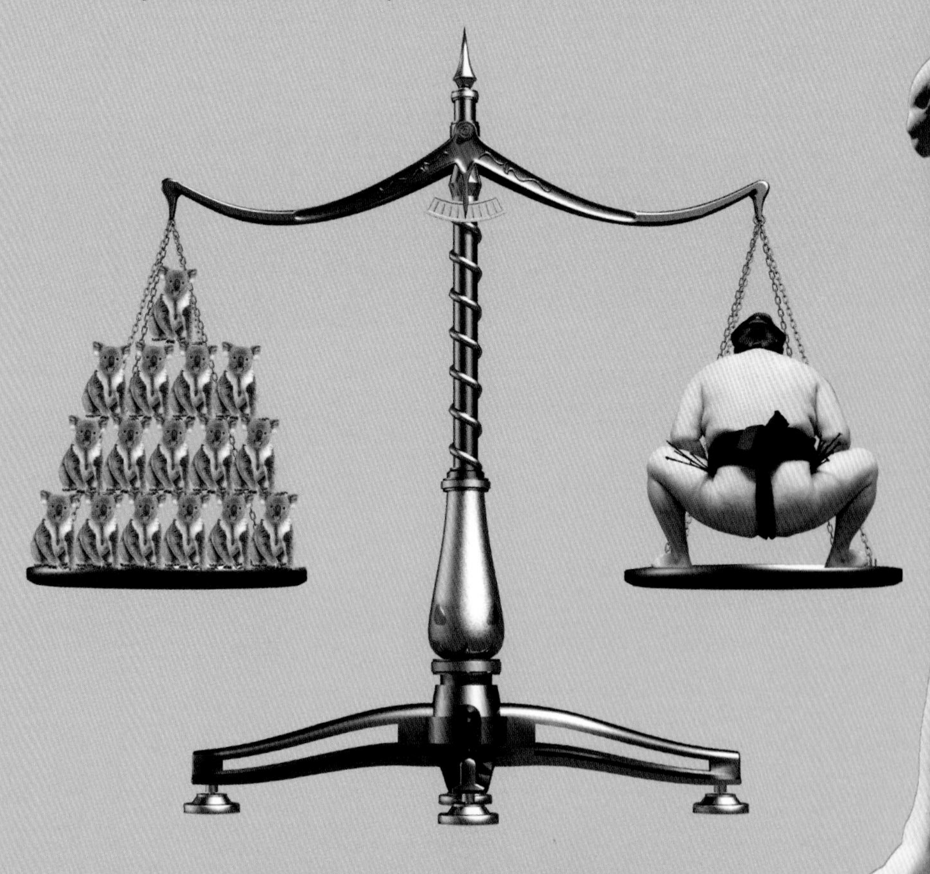

Brain or brawn?

Size isn't everything. It's what you've got up top that counts. Homo sapiens gets his name from his comparatively large brain (about 1.3–1.4kg or 2lb 13oz–3lb 2oz) – but how does he measure up to other species?

Our close relatives, chimpanzees, may be smart, but they don't do so well on the scales. Their brains are only about one-third of the weight of ours, and just a smidgeon bigger than those of cows or donkeys – which are not exactly the brightest stars in the firmament.

The chimpanzee brain, however, represents just under 1 per cent of its body weight: much the same percentage as a pig or horse. And, surprisingly, an African elephant's brain is a massive four times heavier than a human's. Perhaps that's why they never forget.

Compare this with another monster, the diplodocus. This dinosaur's brain was only 0.004 per cent of its enormous 11,700-kg (11.7-ton) body weight, and at around 50g (1¾oz) was just twice the size of a domestic cat's.

Even the wisest bird, the owl – though not exactly a birdbrain – has a brain about 600 times smaller than ours.

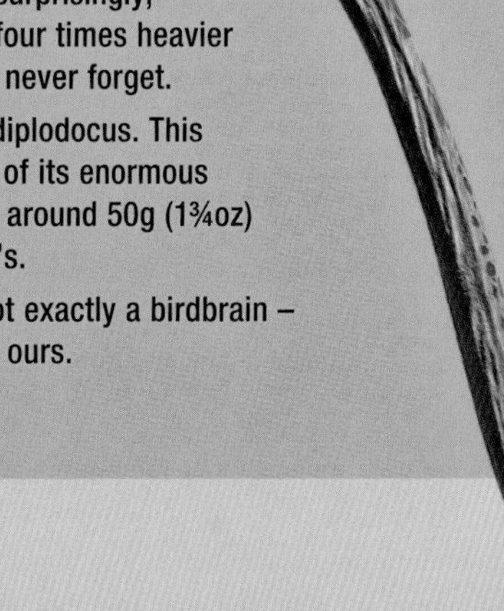

A GRAIN OF TRUTH

Fleas, by their nature, are not easy to weigh. They vary quite a bit in size, too, but an average of all the estimates gives them a weight of roughly one-tenth of a gram (or 0.0035273962oz). Or, more meaningfully, 10 fleas would weigh the same as one push pin. You think that's small? How about a grain of sand? Again, we're talking average here – a medium-sized grain of quartz sand. It would take 10 of them to reach a fleaweight.

1 flea = 10 grains of sand

1 grain of sand = 10–100 million bacteria

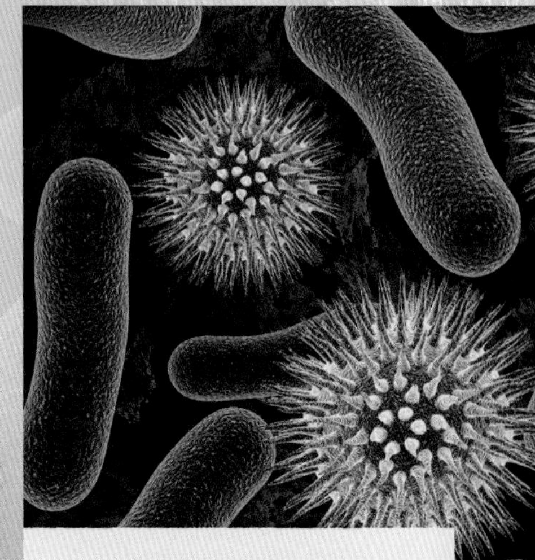

A grain of sand is not very small, in the scheme of things. Compared to bacteria, that grain of sand is a real heavyweight. It weighs anywhere between 10 and 100 million times more than a single bacterium (depending on which type of germ takes your fancy) – and 1,000,000,000,000,000 times an average virus. Now if that virus was scaled up to the mass of that grain of sand, an equivalent grain of sand would weigh 900,000,000,000 tons, give or take the odd million.

1 grain of sand = 1,000,000,000,000,000 viruses

1 virus = 100,000 molecules of water

As light as it gets

When you get down to microscopic levels, the weights are incredibly tiny. So small that scientists argued for some time whether a neutrino, a very small elementary particle, had any mass at all (it has, apparently, but only just).

If we start with a molecule of water, which is about one hundred-thousandth the weight of an average virus, we can go down the scale to some unbelievably small weights. A hydrogen atom is around one-eighteenth the mass of the water molecule it's in, and is more than 1,800 times the mass of any of its electrons.

You just have to wonder how they weigh these things.

WEIGH TO GO

Just as we used the size of a parking space to help us visualise lengths and areas, we can take the weight of an average car as a unit of weight. Now, cars come in a variety of shapes and sizes, from eco-friendly city runabouts to gas-guzzling 4x4s, but if we take a modest family car, it weighs about 25 John Does.

What about the weight of some other vehicles? The double-decker bus, for example? Coincidentally, the double-decker bus weighs much the same as a 10-ton lorry (no prizes for guessing how much), equivalent to 5 family cars, or 125 John Does; whereas a modern Greyhound bus weighs in at a bit more, around 7 family cars, or 2.35 elephants. Compare that with military tanks, which can be as heavy as 20 family cars – almost half a blue whale.

Heavier than air machines

Where elephants and blue whales come into their own is in measuring really heavy things such as passenger aircraft, cruise liners and supertankers. The jumbo jet often gets used as a measure of extremely large weights, especially in the media, but it doesn't mean much – unless we know that we're talking about a fully loaded early model of Boeing 747, which is equivalent to 60 elephants (or 2.4 blue whales). Anyway, for weight, the jumbo has been superseded by that 100-elephant monster, the Airbus 380.

Despite their size, ships are disappointing measures of weight. Probably because they're designed to float, even when fully loaded. The *Titanic* (which floated before the accident) was only about one-third the weight of a blue whale, and modern supertankers, even with a full load, seldom get above 3 blue whales.

A QUESTION OF GRAVITY

Scientists don't like talking about weight. Not because they're sensitive about their physiques, but because the weight of an object varies according to where in the universe it is. They prefer to think in terms of mass.

It's a very difficult distinction for most people to understand – but it's basically a question of gravity.

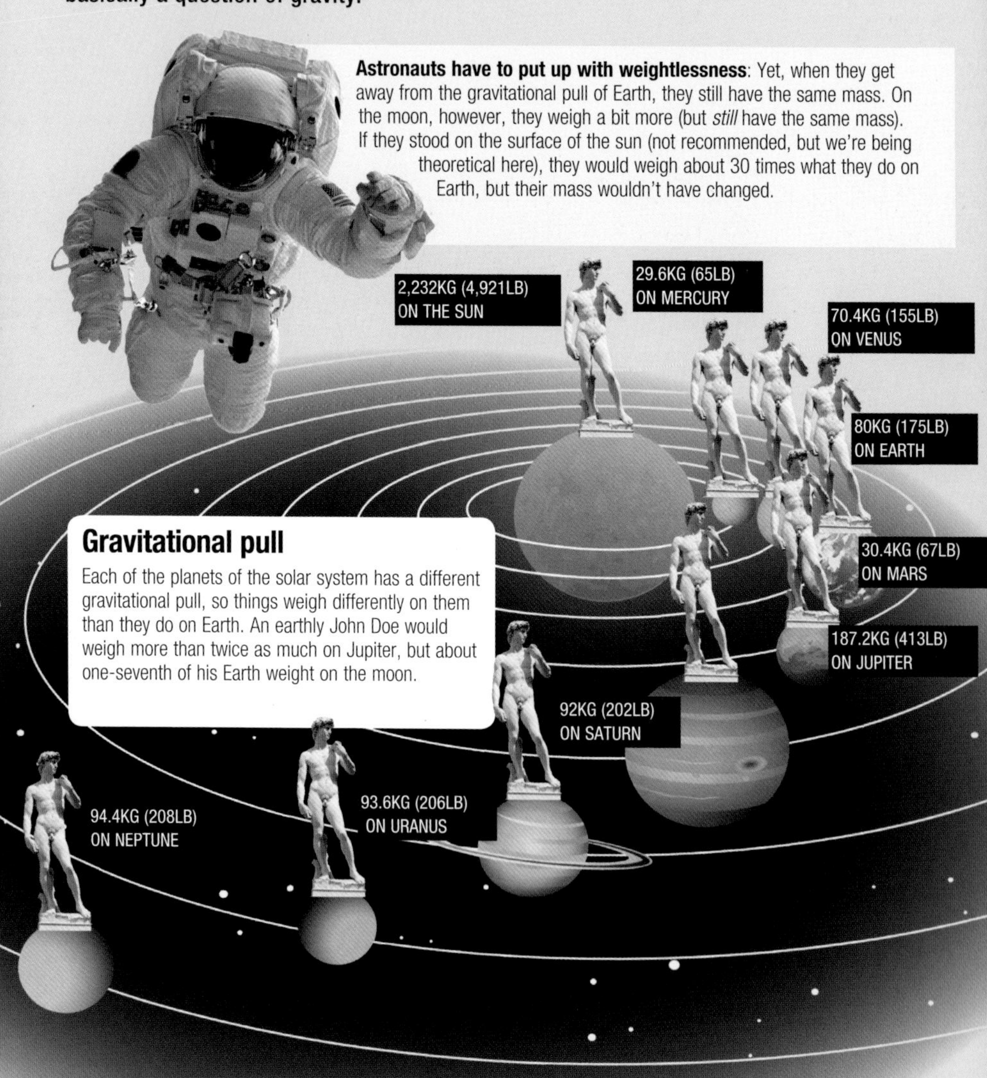

Astronauts have to put up with weightlessness: Yet, when they get away from the gravitational pull of Earth, they still have the same mass. On the moon, however, they weigh a bit more (but *still* have the same mass). If they stood on the surface of the sun (not recommended, but we're being theoretical here), they would weigh about 30 times what they do on Earth, but their mass wouldn't have changed.

Gravitational pull

Each of the planets of the solar system has a different gravitational pull, so things weigh differently on them than they do on Earth. An earthly John Doe would weigh more than twice as much on Jupiter, but about one-seventh of his Earth weight on the moon.

Losing weight

It's not just in space that your weight can vary. The gravitational pull of Earth is weaker at the equator than at the poles by about 0.05 per cent. It doesn't sound much, but a polar John Doe of 80kg (175lb) would weigh about 40g (1½oz) less in Macapá, Brazil. The farther you get away from Earth's surface, the lighter you get, too, by 0.5 per cent for every 10.7km (6mi 3,165ft). On top of Mt Everest, John Doe would weigh about 355g (12½oz) lighter than at sea level, or on a plane cruising at 10,668m (35,000ft), he'd weigh 400g (14oz) less.

WORTH ITS WEIGHT IN GOLD

Just as you can't judge a book by its cover, you can't tell how heavy something is just by looking at it. Size is not a reliable indicator of weight, as any delivery person will tell you: picking up a small packet of lead weights can be quite a shock; yet a large packet of rice cakes seems to weigh nothing at all.

It's all a question of density. Different materials have different densities, which is what makes them seem light or heavy. A pound (or kilogramme equivalent) of cork weighs exactly the same as a pound of gold – a pound – but it takes up more space, because gold is much denser than cork. In fact, a pound of cork would be 77 times bigger than a pound of gold.

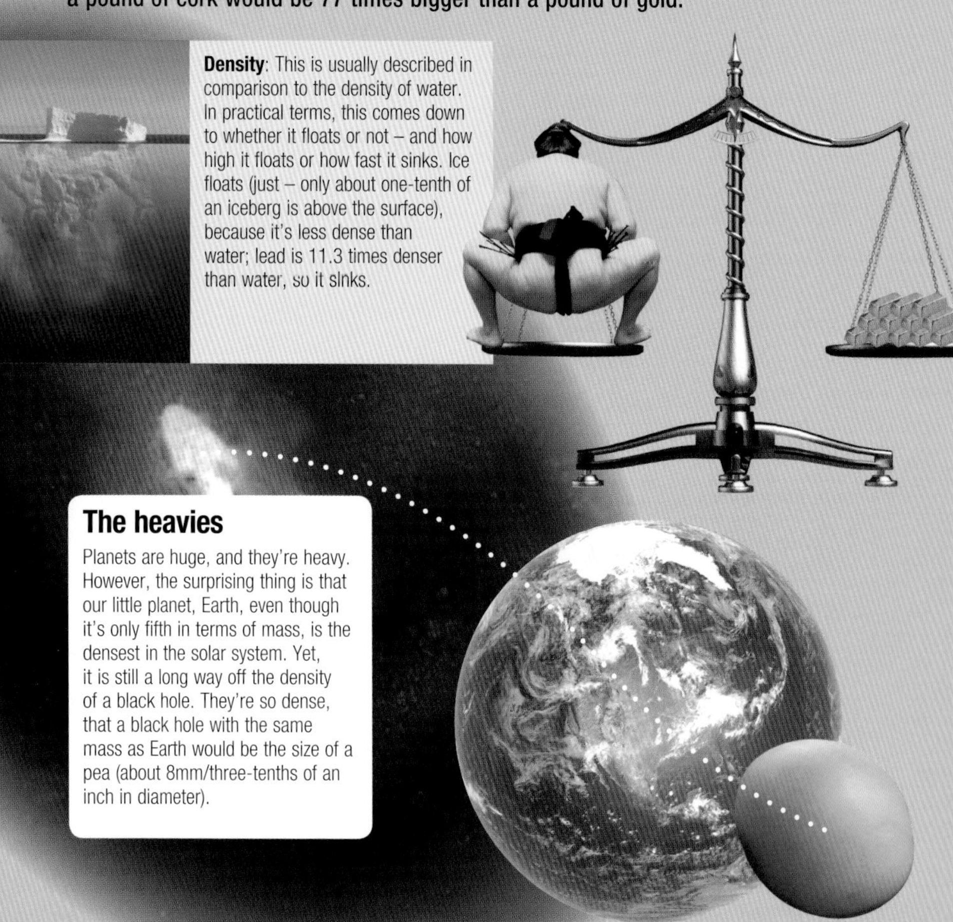

Density: This is usually described in comparison to the density of water. In practical terms, this comes down to whether it floats or not – and how high it floats or how fast it sinks. Ice floats (just – only about one-tenth of an iceberg is above the surface), because it's less dense than water; lead is 11.3 times denser than water, so it sinks.

The heavies

Planets are huge, and they're heavy. However, the surprising thing is that our little planet, Earth, even though it's only fifth in terms of mass, is the densest in the solar system. Yet, it is still a long way off the density of a black hole. They're so dense, that a black hole with the same mass as Earth would be the size of a pea (about 8mm/three-tenths of an inch in diameter).

5 VOLUME & STORAGE CAPACITY

A drop in the ocean: Statistics are often baffling, and statistics about volumes and capacities are more baffling than most – it isn't easy making sense of figures given in cubic metres (m^3). A cubic metre or cubic foot (ft^3) remains a mystery to most people. Pints and litres mean a bit more to us because we're used to those quantities of milk, fizzy drinks and so on. We're a little vague though when it comes to even slightly larger volumes such as how much petrol we're putting in the tanks of our cars. And we're much more hazy about measuring the volume of something like a warehouse, so we need other units to put that sort of thing into perspective.

LIQUID ASSETS

On an everyday level, volume usually means 'how much liquid'? Solid stuff tends to get measured and sold by weight, but we buy our beverages and petrol in pints and litres. Every drinker has a good grasp of what these units mean (although coffee drinkers can be confused by measures such as the Tall, Grande and Venti), so what we know already is a good starting point for thinking about volume.

A 'cup' is a ubiquitous but elastic kitchen measurement: it varies in different countries. A cup is roughly half a pint or a quarter of a litre, which is something we can all relate to. You might prefer to think of it as 2 moderate glasses of wine, assuming you get 6 glasses from a bottle. So let's use these as a way of measuring a bigger volume – a bathtub, say, which would take around 4,000 glasses of wine (Wglass) to fill to the brim, or 666.6 wine bottles (Wbot).

Olympic swimming pool: This measure has gained some currency in the media for very large volumes. It's a nice image, but doesn't tell us much. Until we learn that an Olympic-size pool (Opool) is equivalent to 5,000 bathtubs (Btub). Or 20 million glasses of wine (Wglass). Do the maths.

Room to spare

Volume of liquids is one thing, but converting that kind of measure into the size of rooms, halls and buildings is another. To make the transition use a bathtub as a basic measurement: put four of them together, and you've got the volume of a telephone box (remember that? No? Think Superman). Or an average family saloon car's interior, about 7 bathtubs. From there, it's a short step to sizing up a room; one with a height of 1.5 M. Eiffels, and a floor area of 6 double beds would have a volume of 20 bathtubs, or 5 phone boxes (Pbox). And you could get 250 of those rooms in an Olympic pool.

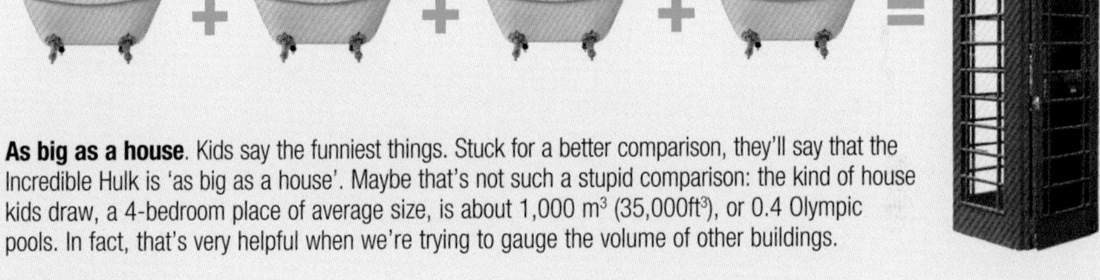

As big as a house. Kids say the funniest things. Stuck for a better comparison, they'll say that the Incredible Hulk is 'as big as a house'. Maybe that's not such a stupid comparison: the kind of house kids draw, a 4-bedroom place of average size, is about 1,000 m^3 (35,000ft^3), or 0.4 Olympic pools. In fact, that's very helpful when we're trying to gauge the volume of other buildings.

The main concourse of Grand Central Terminal in New York for instance; it's got the same volume as about 177.5 Houses. Or the Taj Mahal (200 Houses), which, strangely, is much the same volume as Chartres Cathedral.

Not forgetting that perennial teaser posed by the Beatles: how many holes does it take to fill the Albert Hall? No idea ... but the auditorium takes 100 Houses (see also page 70).

BOXING CLEVER

Thanks to a unusually sensible bout of standardisation, the containers used for hauling and storing freight come in regular sizes. Well, regular enough for the industry to talk in terms of containers without being too wide of the mark. International freight containers come in three different sizes: 6m (20ft), 9m (30ft) and 12m (40ft), all referring to their length. To make matters simpler, the basic unit is the volume of the smallest, known as a 6m (20ft) equivalent unit (T.E.U.) – but for our purposes we'll just call it a container – and it's equivalent to 21.5 phone boxes.

12.5 phone boxes in a container

The capacity of a container ship: This is often quoted in T.E.U., or the number of containers it can hold. This can be as much 10,000, which is about 172 Olympic pools. Put like that, it's easy to compare with the capacity of supertankers, carrying up to 200 Olympic pools of oil.

10,000 containers in a container ship

200 Olympic swimming pools in a supertanker

Conspicuous consumption

When you're filling your car at the petrol station, you watch the figures spin until you've got the amount you want. Because of rising prices, many of us put in 20 pounds' worth or whatever, and have no idea of the volume that represents. Have a look next time you fill up, and picture it in terms of wine bottles. You'd be surprised. An average petrol tank holds about 75 wine bottles.

Multiply that by the number of times you fill up every year, and then by the number of cars on the road, and we're suddenly talking huge numbers. Total consumption of oil is even greater: a family of four in the USA goes through about 275 petrol tanks of oil every year. Just as well oil production is still being measured in millions of barrels (a barrel's just under 3 petrol tanks) – per day.

76 wine bottles in a petrol tank

A DRINK PROBLEM

In much of the developed world, water is taken pretty much for granted. Turn on the tap, and there it is. We speak of spending money like water, like it's an unlimited supply. Even in Australia, the driest continent in the world, the amount of water used is staggering – more than 650 wine bottles of it per person every day. It's higher still in the USA (765Wbot per capita per day).

However, in other parts of the world, it's a more precious resource. Water consumption in China is only one-sixth that of the USA, and the average American's usage is the same as 33 people in parts of Africa.

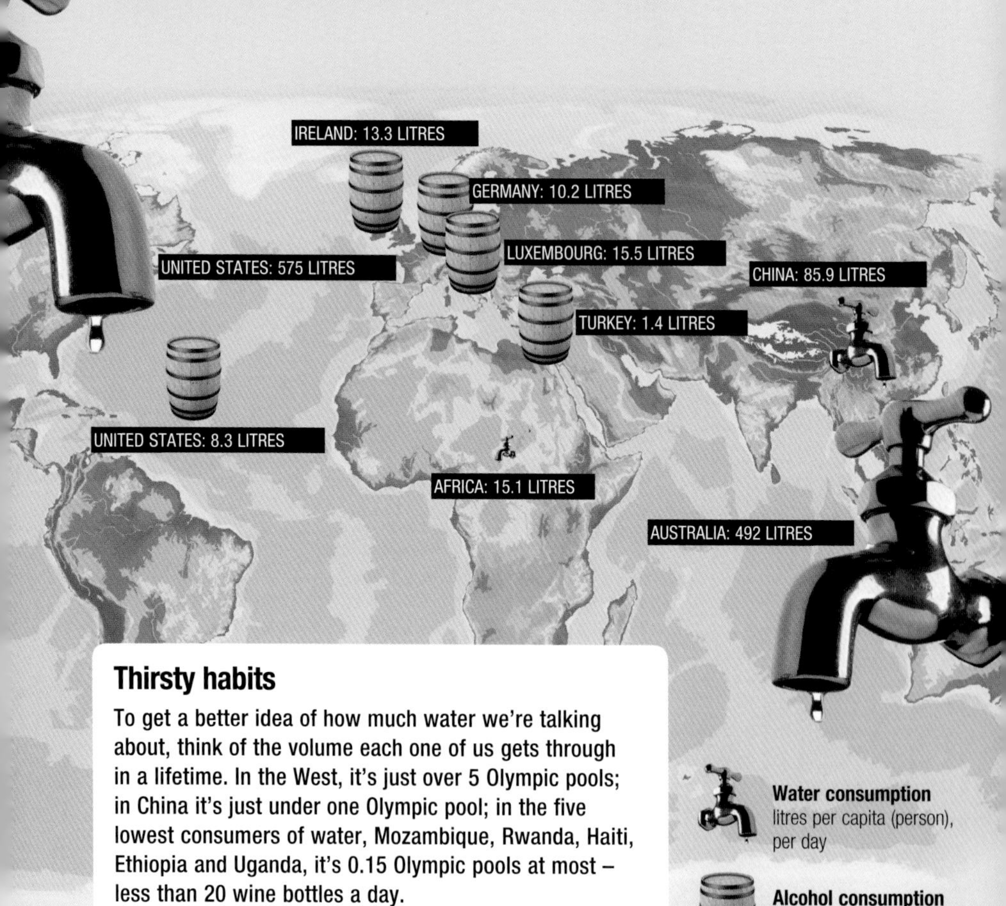

Thirsty habits

To get a better idea of how much water we're talking about, think of the volume each one of us gets through in a lifetime. In the West, it's just over 5 Olympic pools; in China it's just under one Olympic pool; in the five lowest consumers of water, Mozambique, Rwanda, Haiti, Ethiopia and Uganda, it's 0.15 Olympic pools at most – less than 20 wine bottles a day.

Water consumption
litres per capita (person), per day

Alcohol consumption
litres per capita (adult), per annum

Downing it down under

In Australia, they have an ingenious unit for large volumes of liquid: the sydharb – about the volume of water in Sydney Harbour. In practise, this is assumed to be 500 gigalitres (that's 200,000 Olymic pools to you and me).

1 SYDHARB = 500 GIGALITRES (400,000 ACRE-FEET)

The sydharb helps put Australia's consumption of liquids into perspective. They go through 7.65 sydharbs (1,530,442 Olympic pools – let's call that 1.5 million Opools) of water every year. Their legendary capacity for beer-drinking, however, is in fact disappointing – only 935.5 Olympic pools per annum (it would take more than 200 years to down a sydharb).

15,115,151,500,000

beers in Sydney Harbour

Not for the teetotallers

Water's not the only liquid we down in quantity. We drink a surprising amount of alcohol, too – at least, some of us do. As you'd expect, the Middle East and North Africa (predominantly Muslim countries) get through relatively little, but in North America, on average, each adult drinks the equivalent of 11.3Wbot of pure alcohol a year. That doesn't sound a lot, but it's the amount of alcohol in 95 bottles of wine.

The figures are even higher for Europe, where the average adult downs the equivalent of 13.3Wbot of pure alcohol every year, the same as 111 bottles of plonk; and the biggest drinkers of all are apparently the Luxembourgers, who shift the equivalent of 172 bottles of wine every year. Each.

When it comes to beer-drinking though, Luxembourg is way down the list. The prize for ale-swilling goes to the Czech Republic, whose citizens' annual per capita consumption of their favourite export is a mighty 210Wbot – not far short of one-half a bathtub.

WATER, WATER, EVERYWHERE

There's a lot of water on the earth, most of it in the seas and oceans, but also a fair proportion of it in lakes, ponds and rivers. There's so much, in fact, that units such as the Olympic pool really don't help – the numbers still run into the hundreds of millions. Even the sydharb is too small a unit to describe some of the largest bodies of water, and you'll see most figures given in cubic kilometres or cubic miles.

So let's start small. Well, comparatively small. Lake Superior has a volume of about 11,600km^3 (2,800mi^3), or 23,000 sydharbs. Moving up the scale, Lake Baikal in Siberia is roughly twice that volume, and the Caspian Sea is equivalent to the volume of 6.75 Lake Superiors (over 1.5 million sydharbs).

So far, so good. However, with oceans, the statistics read like telephone numbers, even in sydharbs. The Indian Ocean (584,262,000 sydharbs) is better thought of as 3,735 Caspian Seas or 25,000 Lake Superiors; the Atlantic as 300,000 Superiors/4,500 Caspians; and the Pacific as 8,580 Caspians, or very nearly 2 Atlantics. That's much easier to visualise than 1,342,096,000 sydharbs, isn't it!

Niagara Falls = 12,000

Go with the flow

Don't forget the rivers that feed those lakes and oceans. A lot of water flows down them every day. To get some idea of how much, let's take a convenient point on one river where the flow is spectacularly visible: the Niagara Falls. On average, 12,000 bathtubs (6,000m^3 /212,000ft^3) of water tumbles over these falls every second. Keeping the bathroom analogy, that's much the same rate of flow as 43,225,000 showers – or, if you prefer, 2 million toilets continually flushing.

2,000,000 flushing toilets or 43,225,000 showers

Victoria Falls flows at per 1,000m^3 second (38,430ft^3/sec), the equivalent of 333,000 toilets or 7,200,000 showers.

GREAT BALL OF FIRE

Working out the volumes of spheres isn't easy (volume = 4 divided by 3, times πr^3. OK, it's not that difficult, but we're used to doing rough calculations for purposes of comparison), and comparing round objects by volume is very different from looking at their relative diameters or surface areas. Take two balls, one with a diameter three times as big as the other: you'd think it would have three times the volume – but you'd be wrong. If you don't believe it, go away and do the maths.

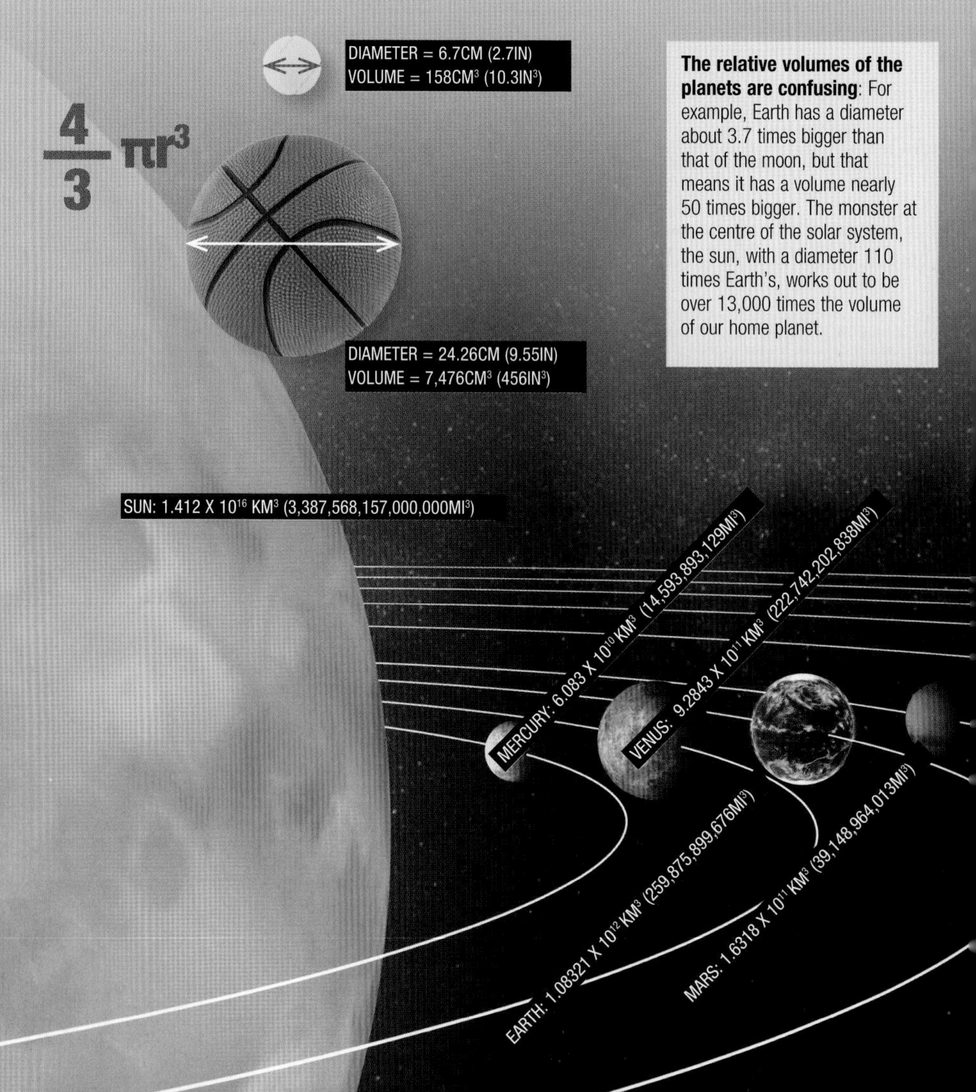

The relative volumes of the planets are confusing: For example, Earth has a diameter about 3.7 times bigger than that of the moon, but that means it has a volume nearly 50 times bigger. The monster at the centre of the solar system, the sun, with a diameter 110 times Earth's, works out to be over 13,000 times the volume of our home planet.

Having a ball

Perhaps it would be easier to imagine the volume of the planets if they weren't round. Take Earth as an example: If we could squash it like a lump of putty into a squarer shape, we'd end up with a cube with sides of 10,270km/6,380mi long – very roughly 1.5 times the length of the Nile. That cube-shaped Earth can also help us to visualise the proportions of the oceans (see previous pages). The Pacific Ocean would fit in a cube-shape container with sides of 875km (544mi); put that inside the Earth cube and you can see just how little that is compared with the volume of Earth, despite covering a good proportion of its surface.

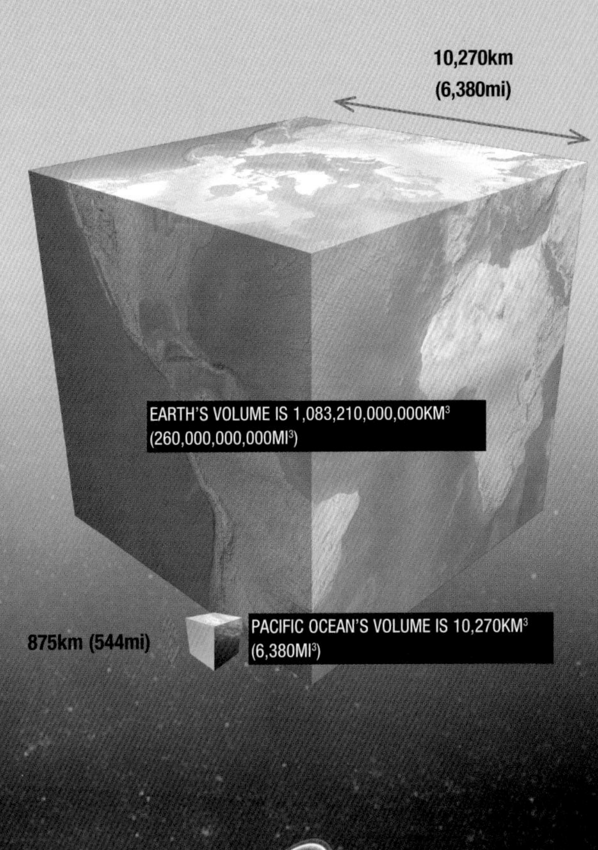

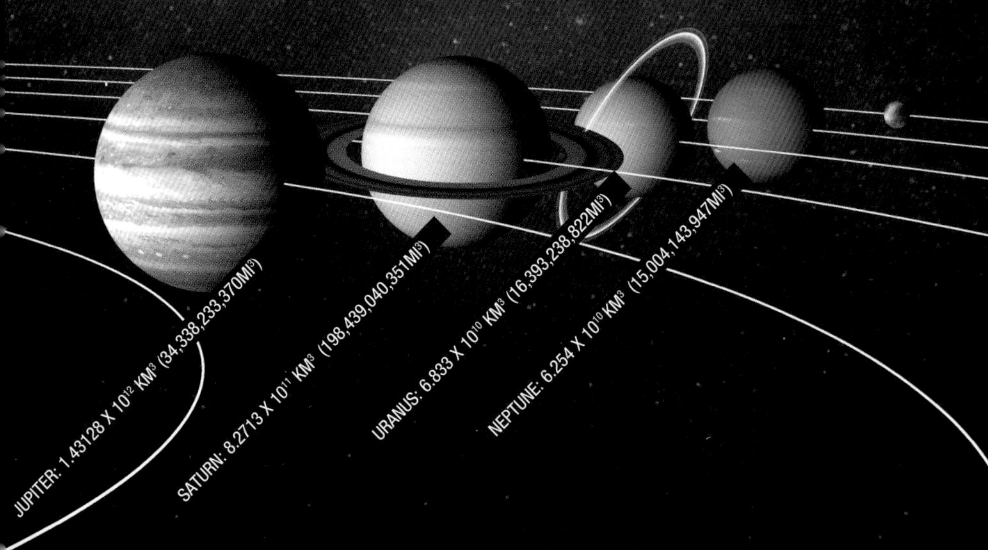

FOR YOUR INFORMATION

Now that owning a bunch of electronic gadgetry is almost universal, we're used to dealing with units of memory, but we're so blasé about it we've lost touch with what things like megabytes and gigabytes actually mean.

Before computers, information was stored in books – and a book could hold a lot of information. The benchmark, not exactly the sum of all human knowledge but a good proportion of it, was the *Encyclopaedia Britannica*; put into electronic terms the text would be the equivalent of around 300MB. You could fit two copies comfortably on a CD-ROM. Or 150 copies of the King James Bible, if we take a Bible (Bib) to be about 4.3MB (much the same as a track of music in mp3 format).

Thanks for the memory. Estimates for the capacity of the human brain vary wildly, depending on how the figure is arrived at – do you count just neurons, or synapses? Whatever, the figures are huge. Somewhere between 10 and 1,000 terabytes, if the neuroscientists are to be believed, or the equivalent of between 1 and 100 Libraries of Congress. Accessing information from such a huge store is another question. That's why you can never remember your mobile phone number.

What's the capacity of a DVD? Amazingly, it's about 15 *Encyclopaedia Britannicas* (EBr) – which would take up a heck of a lot of shelf space in a library (a CD-ROM can hold 2 EBr). And whilst we're talking about libraries, it's been estimated that the entire contents of the Library of Congress (excluding pictures) works out at 10 terabytes, which could be stored on 2,128 DVDs.

Next time you turn on your PC, check the size of its hard disk. If it's an average size (these days, anyway) 250GB, that's roughly 800EBr, or getting on for 60,000Bib. Even your 4GB iPod can manage 12EBr. And it would only need 40 PCs (or 2,500 iPods) to contain one Library of Congress.

=

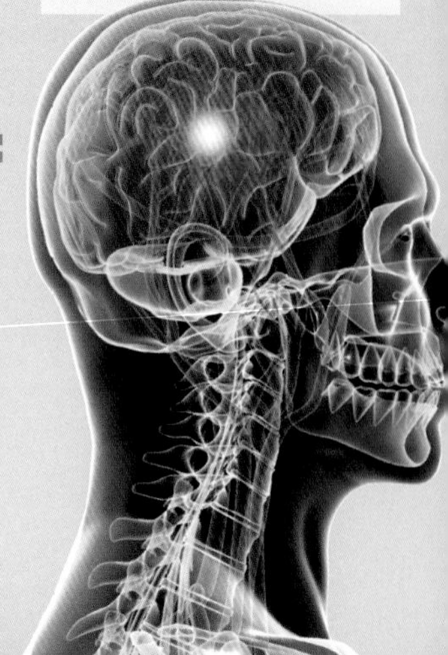

6 POPULATION

One in a million: It's not easy to estimate the size of large crowds of people – ask after any protest march how many took part and you'll get vastly different answers from the protesters, police and those against the protest. We're just as vague about the population figures of our towns and cities; and when it comes to countries such as India or China, the statistics have way too many zeroes for our brains to take in. As with all huge numbers though, we can break the figures down into comprehensible chunks.

1,335,960,000

A CAPTIVE AUDIENCE

Remember the last time you went to a show (or the films, a concert or whatever)? How many people were in the audience? Unless there were only a dozen or so, it isn't easy to guess. If our ability to gauge numbers at that level is bad, how can we begin to think about bigger figures?

The answer, of course, is to break them down into manageable units; things we can see in our mind's eye. Like that old favourite, the Greyhound bus. A typical bus carries around 50 people, so divide the audience into bus-sized groups and it's suddenly easier to estimate the total.

50 people in a Greyhound bus

10 busloads in an Airbus 380

Bus capacity

Using the bus analogy, you can get a good idea of the capacity of aircraft such as the jumbo and Airbus 380 (around 10 busloads), and even places such as Carnegie Hall (56 busloads), the Albert Hall (100 busloads and the Hollywood Bowl (about 30 Airbusloads).

100 busloads in the Albert Hall

Living together

It's one thing gauging the size of a crowd in one place, quite another making sense of things such as the population statistics of towns and cities. Easy enough if you live in a small town of about 1,000 souls (that's 20 busloads), but what about places like New York or London?

Again, it helps to break things down. Think of the passengers waiting to board an Airbus, and you've got 500 in your mind. Just work up from there and you can visualise towns of 100,000 (200 Airbuses), and move on up to cities such as Odessa, with a population of a million, then Beijing (population *c.*10 million). Just maybe, it becomes easier to comprehend.

Antisocial?

We aren't the only animals that gather in large social units. Although most species form small family groups, some, especially insects, live in colonies of hundreds, thousands and even millions – just like our towns and cities. Termite and ant colonies can interconnect, too, in supercolonies of billions that are analogous to our countries. They're in anthills and underground, so we're not aware of them.

More noticeable (or spectacular) are locust swarms. These can cover hundreds of square kilometres/miles in area, and contain tens of billions of insects – many times the human population of the world.

60,000,000,000
locusts in a swarm

WHAT A STATE

There are a lot of people in the world. More than 6.8 billion, and rising. Obviously we can't think of that in terms of busloads, or even Hollywood Bowls. Luckily, we humans have divided the planet into more manageable units: countries.

Unfortunately, they're far from uniform in size. Some of them such as the Falkland Islands (pop. 3,000) are no bigger than towns, while others are a significant percentage of the world population, and somewhere in the middle we have countries such as France, Italy, the UK and Thailand (all with a population of 60–65 million each). Let's take Burma (population of *c.*50,000,000) as a benchmark. This is the equivalent of 5 Beijings (or 6.5 Londons, or 7 New Yorks), and almost exactly half the population of Mexico.

Percentage of world population

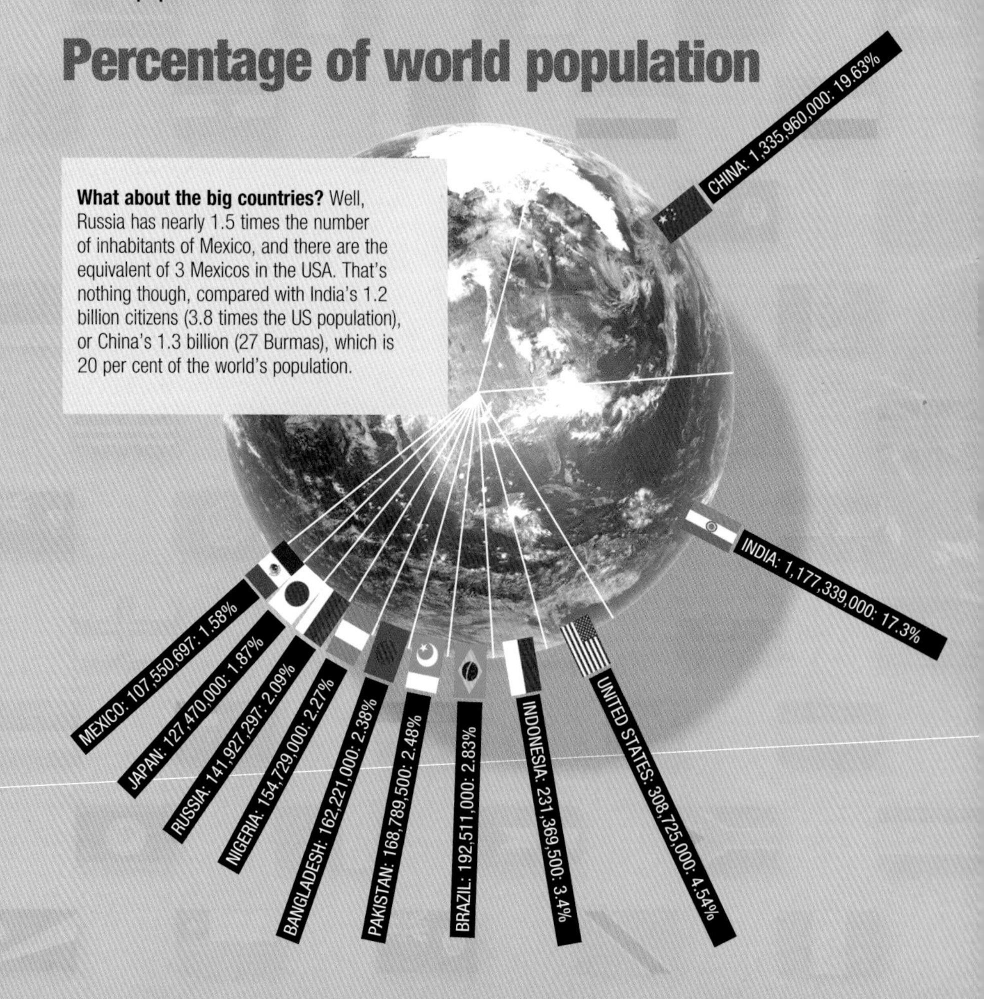

What about the big countries? Well, Russia has nearly 1.5 times the number of inhabitants of Mexico, and there are the equivalent of 3 Mexicos in the USA. That's nothing though, compared with India's 1.2 billion citizens (3.8 times the US population), or China's 1.3 billion (27 Burmas), which is 20 per cent of the world's population.

A growing number of people

One of the problems with measuring populations is that they change so rapidly as people are born and die. Overall, populations are continuing to grow, and the world seems to be becoming a more and more crowded place. Only 2,000 years ago, there were about as many people in the whole world as live in Brazil today, and 1,000 years ago that number had only risen by half as much again.

By 1800 we had reached about one-fifth of our present population, and from then on the rise has been increasing: in the last 50 years, the world population has more than doubled. Even assuming a slowdown in growth, predictions are startling.

UNITED KINGDOM: 1.82

UNITED STATES: 2.05

GUINEA-BISSAU: 7.07

INDIA: 2.81

HONG KONG: 0.97

Key: Fertility rate 2005–2010 (births per woman)

A matter of life and death

Population growth is complicated to predict, because it's affected by a number of factors. People are living longer, and more children are surviving to sexual maturity and producing children, pushing the figures up; on the other hand, in the richer countries at least, the birth rate is falling. Poorer countries, where infant mortality is high, tend to have a much higher birth rate.

EXTENDED FAMILIES

Humans tend to produce their offspring one at a time, although twins and triplets are not uncommon. Multiple births, up to eight or nine, are rare and often the result of assisted reproductive technology. Other mammals have litters of up to 14, birds lay clutches of 1–15, and reptiles up to a few dozen, but amphibians, insects and fish don't take any chances. They can lay thousands, or in some cases millions, of eggs each season. The queen termite wins the prize for best mum though – her life is devoted to producing thousands of eggs every day.

5,000+ termite eggs a day

Time flies

Multiple births in the animal kingdom are, thank goodness, countered by low life expectancy. If they weren't, we'd soon be overrun by all kinds of creatures. Houseflies, for instance, reach maturity in a matter of days, lay eggs in batches of hundreds and can live as long as 3 weeks. If this wasn't checked by a severe mortality rate, one pregnant female could engender a dynasty the size of the world's human population in around 5 weeks.

It takes all sorts

If you thought humans ruled the world, think again. There's only 6 billion of us, and we're just one of the 5,500 species of mammals on the planet. As far as we know, that is – there may well be more. Scientists have so far managed to identify about 1.5 million different species of animals, but it's likely only the tip of the iceberg. It's possible that there are as many as 10 million species of insects alone, outnumbering all the others by 10 to 1. That puts us in our place.

20 per cent of species extinct by 2040

Dead as a dodo

Almost as fast as we discover new species, established ones are disappearing. At the current rate of extinction (that we know about, anyway), it is estimated that up to 20 per cent of species will have disappeared in the next 30 years, and pessimistic forecasts predict the loss of half of them in the next century. Which ones? Who knows?

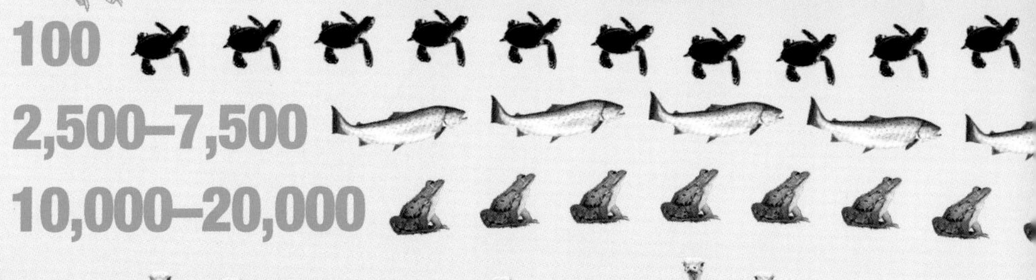

COMPARATIVELY SPEAKING

As modern technology makes communication easier, making yourself understood has become more of an issue. There are thousands of languages spoken across the world, and most of them are mutually unintelligible. The languages with the most speakers are growing and spreading like empires, swallowing up the smaller ones, and the need to communicate means increasing numbers of people speak one or more of the major languages – maybe one day settling on a 'lingua franca'. Going by current figures, Mandarin Chinese is the most likely candidate.

Fewer and fewer speak the little languages, which are dying out at a more alarming rate than animal species. Some, however, are enjoying a revival, especially in areas with a fiercely nationalistic culture or politics. No sign of Welsh or Basque disappearing, for example, even though they are restricted to populations smaller than many cities.

Believe it or not: Just like languages, religions are broadly speaking regional in their distribution, but migration and evangelism have spread the major faiths worldwide. Similar to languages, it's difficult trying to put numbers on the different beliefs. In Christian countries, for instance, many non-believers are classified as Christian and even say they are so in official forms (in a recent UK census, a significant number of people claimed Jedi as their religious belief – you can't believe everything people say). Similarly, many people would identify themselves as Jewish, but use the word as a cultural or ethnic rather than religious definition.

Speaking in tongues

It's difficult to get an accurate number for the native speakers of any language, but even more difficult to estimate how many people can speak or understand it as a second (or third, or fourth) language. The numbers are huge for the top few tongues, which together include a good proportion of the world.

If you can speak a couple or so of the major languages (Mandarin Chinese and English would be a good combination), you'd be able to communicate with nearly one-quarter of the world's population; that's roughly 1,700,000,000 people. Either of those languages would be a better option as a second language than Esperanto, the artificial international language that never caught on. You'd be able to chat away to about 2 million people, many of whom already speak another of the top 10 languages anyway.

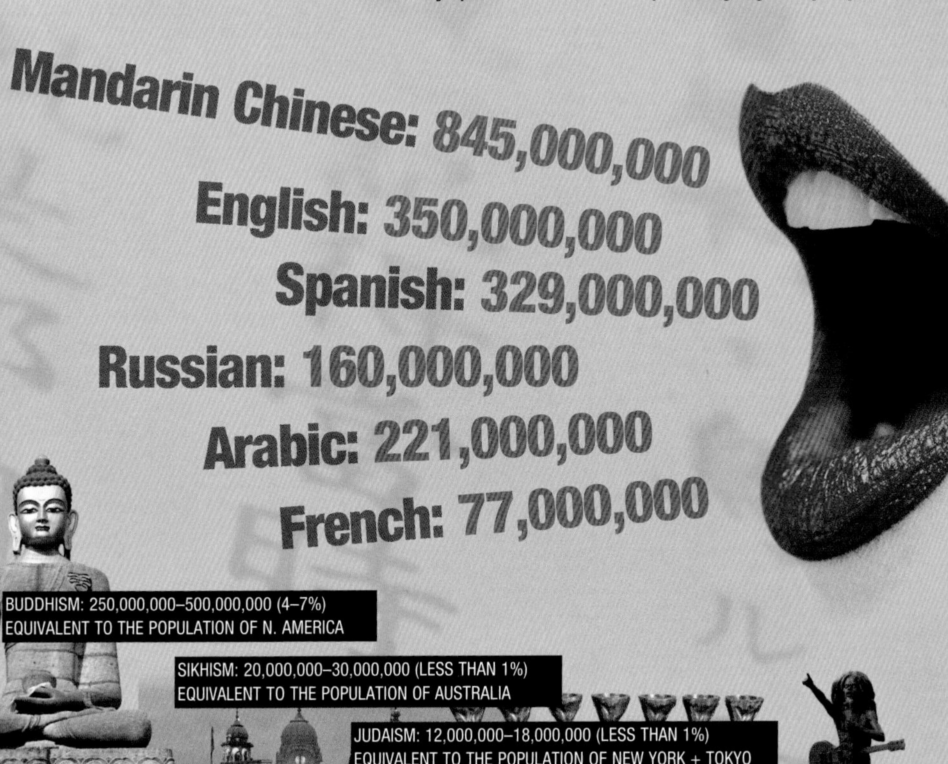

CLOSE NEIGHBOURS

The world's a big place – big enough for all of us to have a fair amount of personal space. If we take just the land area of the earth, there's about 4 American football fields (see Chapter 2: Area) available for each of us, yet we tend to huddle together in cities, leaving vast areas of the world practically deserted. However, places such as Greenland and the Western Sahara are pretty inhospitable, but perhaps even they are preferable to the 1.5 boxing rings each resident of Macau has to live in.

4 football fields of personal space

Getting away from it all. We've largely abandoned our nomadic lifestyle, but there is still a significant movement of populations, nowadays for economic or political reasons rather than chasing food. Of course, there is also the annual migration to holiday destinations. There are, however, animals that make annual journeys, sometimes of thousands of kilometres (or miles), in huge numbers. Penguins trudge across Antarctica to their nesting grounds, and migratory birds in their millions fly from hemisphere to hemisphere to avoid the winter.

Perhaps the most spectacular is the 'Great Migration' from the Southern Serengeti to the northern edge of the Masai Mara in Africa. More than a million wildebeest (accompanied by a third of a million gazelle, 200,000 zebra and many thousands of eland) make a 500-km (300-mi) return trip each year – equivalent to the population of Barcelona (and their pets) travelling around Europe simultaneously.

1,300,000 wildebeest, 360,000 gazelle, 191,000 zebra and 12,000 eland

TIME

Time is a relative concept: At least, so Einstein would have us believe. Sometimes it passes all too quickly, at other times it seems to drag interminably. Even our basic units of time, days, are somewhat elastic – think of the long, hot days of summer, or the long, dark nights of winter. Consequently, the words we use to describe time are correspondingly fuzzy, and not a lot of use for measuring it. There are, however, some constants (or near constants) that we can use. Astronomical, natural and even subatomic cycles provide us with some units that are surprisingly user-friendly.

WHAT A DIFFERENCE A DAY MAKES

The most basic divisions of time for us are the day, the period between two midnights, and the year, the period from one January to the next. They're so much a part of our lives that we take them for granted, and don't even think about what they actually are.

A day (24 hours, in our way of dividing our time) is the time it takes for Earth to rotate once on its axis; a year the time it takes for Earth to complete one orbit of the sun. They're constant (for the time being, at least), so are a fairly reliable way of measuring time. That is, unless you're somewhere other than on this planet.

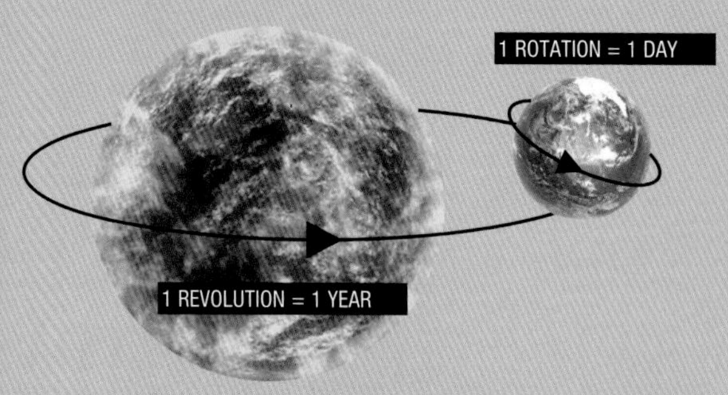

PLUTO: 1 DAY = 6.39 DAYS
1 YEAR = 248.59 YEARS

VENUS: 1 DAY = 243 DAYS
1 YEAR = 224.7 DAYS (OR 0.61 YEAR)

365.242 days = 1 year

Elsewhere in the solar system: Things elsewhere are very different. Pluto, for instance, rotates a bit more slowly, and takes a much longer time to make its way around the sun. If you lived there, your day would be nearly a week long, and you'd have to wait nearly 250 years for each of your birthdays. And on Venus, clocks and calendars would be mighty complicated – a Venus day (243 Earth days) is actually longer than a Venus year (224.7 Earth days).

Marking time: You'd think it would be a simple matter, given two immutable units like the day and the year, to divide time into useful chunks. But you'd be wrong. Dividing up the day's not too much of a problem – we've developed a rather eccentric system of 24 hours, divided into 60 minutes and 60 seconds. But the year is much more awkward.

In ancient Egypt, they thought, logically but wrongly, that a year was 360 days long, and consequently had to keep adjusting their calendars to fit. It was the same with the Chinese. After a lot of refinement, a 365-day calendar evolved, but even this doesn't account for the fact that a year is more like 365.242 days. We're still having to adjust our calendars with leap years and even leap seconds to make it fit reality.

Take a look at the irregular month-lengths in the modern calendar – a real botch job. And a 7-day week? But the Islamic and Hindu calendars, based on lunar months, fare no better: at 29.53059 days, the time between full moons is not a convenient unit either.

The generation gap

For periods of more than a year, we've come up with the decade and the century. All conveniently decimal, but not particularly user-friendly. In conversation, people talk of lifetimes and generations: more human, and nicely vague depending on where and when you're thinking about. To make them into useful units let's extend the biblical three-score years and 10 to 75 for a lifetime, and narrow the child-producing age down to give us a 25-year generation. Roughly speaking, of course.

1980: INVENTION OF COMPACT DISC

1969: NEIL ARMSTRONG AND BUZZ ALDRIN ARE THE FIRST MEN ON THE MOON

1564: WILLIAM SHAKESPEARE BORN

1712: THOMAS NEWCOMEN INVENTS STEAM PISTON ENGINE

1792–1815: NAPOLEONIC WARS BETWEEN FRANCE, RUSSIA, BRITAIN, PRUSSIA AND SWEDEN

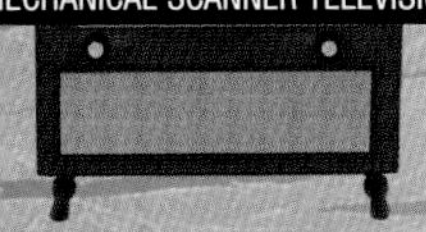

1926: JOHN LOGIE BAIRD INVENTS MECHANICAL SCANNER TELEVISION

1876: ALEXANDER GRAHAM BELL PATENTS TELEPHONE

1512: MICHELANGELO COMPLETES SISTINE CHAPEL FRESCOES

1510 1560 1610 1660 1710 1760 1810 1860 1910 1960 2010

THE TIME OF YOUR LIFE

For most animals, life is nasty, brutish and short. From birth (or hatching), the main point is to grow to sexual maturity, procreate and die. If you're of a gloomy disposition, that just about sums up human existence, too. Or you could take a more optimistic view and divide life into stages. The consensus is that there are four ages of man – infancy, adolescence, adulthood and old age – but some divide it differently, and Shakespeare even gives us 7.

The 7 stages of life

All the world's a stage,
And all the men and women merely players:
They have their exits and their entrances;
And one man in his time plays many parts,
His acts being seven ages. At first the infant,
Mewling and puking in the nurse's arms.
Then the whining schoolboy, with his satchel,
And shining morning face, creeping like snail
Unwillingly to school. And then the lover,
Sighing like furnace, with a woeful ballad
Made to his mistress' eyebrow. Then a soldier,
Full of strange oaths, and bearded like the pard,
Jealous in honour, sudden and quick in quarrel,
Seeking the bubble reputation
Even in the cannon's mouth. And then the justice
In fair round belly with good capon lin'd,
With eyes severe, and beard of formal cut,
Full of wise saws and modern instances;
And so he plays his part. The sixth age shifts
Into the lean and slipper'd pantaloon,
With spectacles on nose and pouch on side,
His youthful hose, well sav'd, a world too wide
For his shrunk shank: and his big manly voice,
Turning again towards childish treble, pipes
And whistles in his sound. Last scene of all,
That ends this strange eventful history,
Is second childishness and mere oblivion,
Sans teeth, sans eyes, sans taste, sans everything.

However you slice the cake, we get a better deal than most other mammals. They generally don't have the luxury of old age, thanks to predators and disease, and their childhood is much shorter – many animals can walk within hours of birth. After that, there's little time for adolescent fun; once they have reached sexual maturity, they settle down to start a family.

Great expectations

In addition to our lifetime, there's a brief period of life before we're born. From conception to birth is roughly 9 months (between 260 and 295 days) for us humans, about 1 per cent of a lifetime. As we tend to produce one child at a time, we reproduce comparatively slowly, unlike rabbits, with a gestation period of around 33 days and litters of up to a dozen – but then they live to only about 10 years old. Elephants, however, live to a ripe old 60 years, and spend about 3 per cent of their lifetimes in the womb.

Average gestation periods

Metamorphosis

The growing-up process in mammals is far less dramatic than it is for insects. They undergo physical changes so drastic they can be completely different creatures at various stages of their lives. After hatching from an egg, a typical insect such as a mayfly goes through a larval stage, then turns into an inactive pupa, before emerging as the insect we recognise. Some insects only live for a few weeks or even days in their adult breeding form, and many of them spend years as larvae preparing for it. In human terms that would be like having an almost lifelong childhood (eating, mainly), a brief, somnolent adolescence and then only a few years adulthood (just enough to produce a family) – and no old age. Maybe not such a bad life ...

A DAY IN THE LIFE

One way of putting a lifetime into context is by scaling it down to a manageable size, such as a day, or a year. If our average European or American lived only as long as our hypothetical mayfly (about 16 hours instead of 80 years) his life would go something like this:

Born at 8.00 a.m., he starts walking and talking around breakfast time, and goes to school at about 9.15 a.m. He has his driving lessons mid-morning, and goes to college at 11.40 a.m. He graduates just before lunch, and has a busy midday finding a job, getting married and settling down to his career. His three kids are born by 2.30 p.m., meaning he spends most of the afternoon as the breadwinner of the family, getting a deserved promotion at around 4.00 p.m. The children leave home and get married between 5.30 and 6.00 p.m., and he takes life a little easier until he retires at 9.00 p.m., getting most of the evening to do pretty much what he likes. He is taken ill just after 11.00 p.m., and moves into a residential home, but dies on the stroke of midnight.

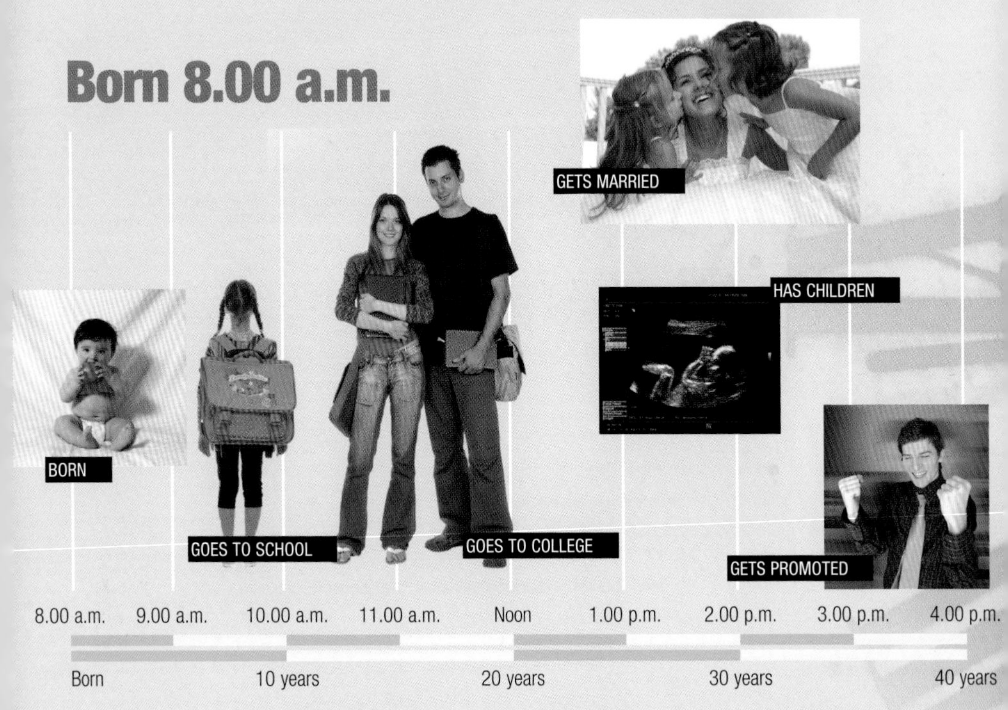

It's a dog's life

How long is a lifetime? Well, of course, that depends … Are we talking human, or dog, or what? And where?

In Europe and North America, people are living longer than ever before – the average life expectancy is up to about 80 years. However, in parts of Africa, it can be as low as 40. So for convenience sake, let's take a round number and call a lifetime 75 years (this also fits in nicely with the generation of 25 years (see page 81).

So how does that compare with other animals? Well, the longest-lived is probably the giant tortoise, which regularly reaches 150 years (2 lifetimes), and the shortest-lived is the gastrotrich (a minute aquatic animal), which gives up the ghost after only 3 days. The old adage about a dog year being equivalent to 7 human years is not far from the truth: dogs live about 12 years on average, so a lifetime is about 6.5 doglives. However, the saying about the mayfly only living for a day isn't true – that's just its adult stage (which may in fact last for several days), after many months or even years underwater as a nymph.

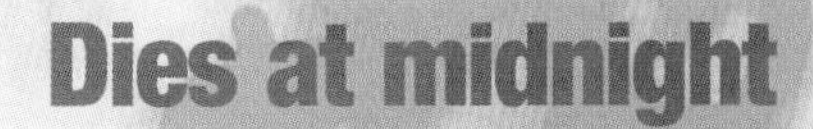

AS OLD AS THE HILLS

When it comes to looking at the lifetime of the earth, we're going to need some big units of time. Some really big units. Where history is divided into centuries and millennia, geological time is usually dealt with in millions of years – megaannums (Ma). Even so, we'll have to deal with some hefty numbers, because the earth was formed around 4,600Ma ago, and life started here about 4,000Ma ago.

If you find that difficult to imagine, try this. Compress the whole of the life of the earth into a single year: it was formed on 1 January, and the present day is the stroke of midnight on December 31. On this geological calendar, the moon appeared on 6 January, and life of some type began halfway through February. Other than that, pretty much nothing happened until April, when primitive photosynthesis got going, and it wasn't until September that any multicellular life developed. The first animals didn't appear until well into November, and the dinosaurs ruled the earth from about 10–20 December. Finally, the first recognisable humans came on the scene at around 8.00 p.m. on New Year's Eve.

251 million years ago

ANIMAL LIFE

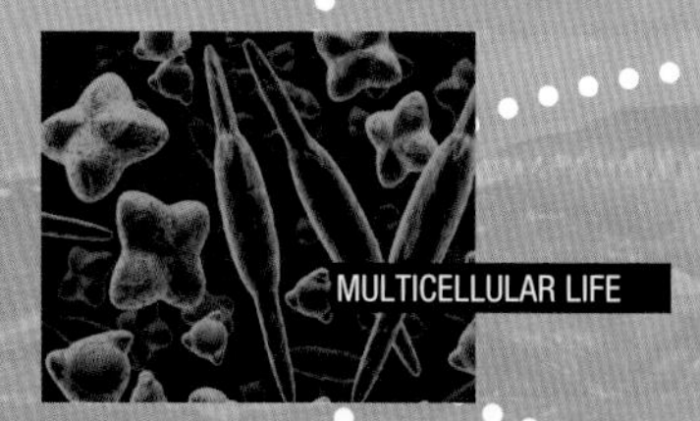

MULTICELLULAR LIFE

PRIMEVAL SOUP

13 billion years ago

Earth is an estimated 20G.Y. old

The Galactic Year

Without going into the mind-boggling idea of when time began, or the question of why there was no time before then, we know that the universe has been around a long, long time. Since the Big Bang, apparently, and that was around 13.3 to 13.9 billion years ago. Probably.

That's not a figure that's easily broken down into centuries, millennia or even megaannums. So how about using the time it takes for our solar system to orbit the galaxy as a unit? It's approximately 250 million years, and is known as the Galactic Year (G.Y.). That puts the Big Bang at a more manageable 55G.Y. ago (or thereabouts). Earth was formed some 35G.Y. later, and life appeared on it about 16G.Y. ago. Now doesn't that make it easier to understand!

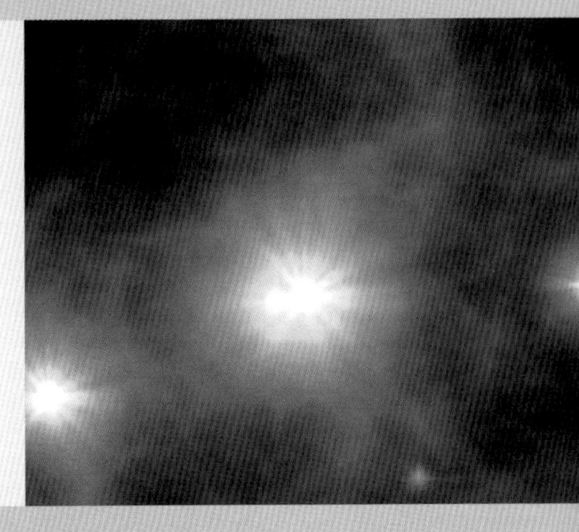

65 million years ago

DINOSAURS

MAN

AS QUICK AS A FLASH

In everyday usage, short units of time are conveniently imprecise: just a moment; it'll take a jiffy; in two shakes, etc. Even the precise measurement of a second is somewhat elastic in phrases such as 'wait a second'. So when we want to describe some really short periods of time, we're stuck for any kind of comparison.

There is, as you would expect, a scientific definition of a second: it is in fact the duration of 9,192,631,770 periods of the radiation corresponding to the transition between the two hyperfine levels of the ground state of the caesium 133 atom, if that's any help. It can be divided into smaller units such as nanoseconds (millionths of a second). Let's be honest, that's no good to most of us.

More practically useful, however, are units we can relate to such as a photographer's flash, for instance. This is typically between 0.001 and 0.005 seconds in duration. For convenience, let's say one-thousandth of a second with modern equipment. So, 'as quick as a flash' gains a precise meaning, much more accurate than a snap of the fingers or the blink of an eye.

0.001 seconds = a flash

Never let it be said, that scientists and engineers have no sense of humour though. They've come up with a couple of nicely whimsical units: the jiffy (0.01 seconds), and the shake (10 nanoseconds).

10 nanoseconds = a shake

0.01 seconds = a jiffy

Microfortnight

Some subversive geeks with time on their hands have come up with a whole different system of units, the FFF system, based on the old measures furlong, firkin and fortnight. From the basic unit of the fortnight (which is 14 days), they derived the facetious microfortnight, one-millionth of a fortnight, or 1.2096 seconds. Very helpful.

8 SPEED

As long as it takes: It's an increasingly fast world we're living in. Communications have sped up, transport is quicker than ever before, speed records continue to be broken and there's even fast food. We expect things to happen as quick as a flash, or faster than greased lightning. Yet we have little real concept of the speeds we're dealing with. A long-haul plane journey seems just that, a long haul; but it's the fastest most people ever achieve. Only a couple of lifetimes ago, speeds like that would have been unthinkable; now we don't give them a second thought. Perhaps we should, and see how they measure up to speeds in the natural world. You'd be surprised.

THE PACE OF LIFE

As with so many things, to get some perspective of speed it's necessary to find a point of reference; something that we can visualise. Walking is the speed we know best, and a brisk walking pace of about 6 kilometres per hour (6kph) or 3¾ miles per hour (3.75mph) is as good a benchmark as any. You can appreciate a good sprinter who's running six times as fast as that.

Animal speeds put our human efforts in the shade. The cheetah is the undisputed champion, over short distances at least, capable of a dash three times as fast as a human sprinter, and enough to rouse a traffic cop's suspicions on the motorway. Race-horses would beat us hands down, too, covering the course in half the time a human runner would take. There are even fish and insects that outstrip our best runners.

It is some consolation then that we can, just like the hare, outrun a tortoise. And, of course, that legendary slowpoke, the snail, is 450 times slower than our brisk walk.

Growing up

Don't let the grass grow under your feet. Actually, you'd have to be a real slow mover to let that happen. However, there's a difference between moving and growing rates. We move comparatively fast, but grow slowly; plants, on the other hand, hardly move at all, but grow relatively fast.

Not all are swift growers, it has to be said – most trees take years to reach their mature heights – but there are some that you can almost see growing. Plants (and weeds) can appear apparently overnight in the garden, and left alone can soon take over. And yes, grass does grow quickly. Some species can grow as much as 15cm (6in) a day. It's the bamboos that take the record. Some bamboo plants can reach 76cm (30in) or more in just 24 hours. Humans take about 2.5 years to grow that much in height.

BAMBOO CAN GROW 76CM (30IN) A DAY

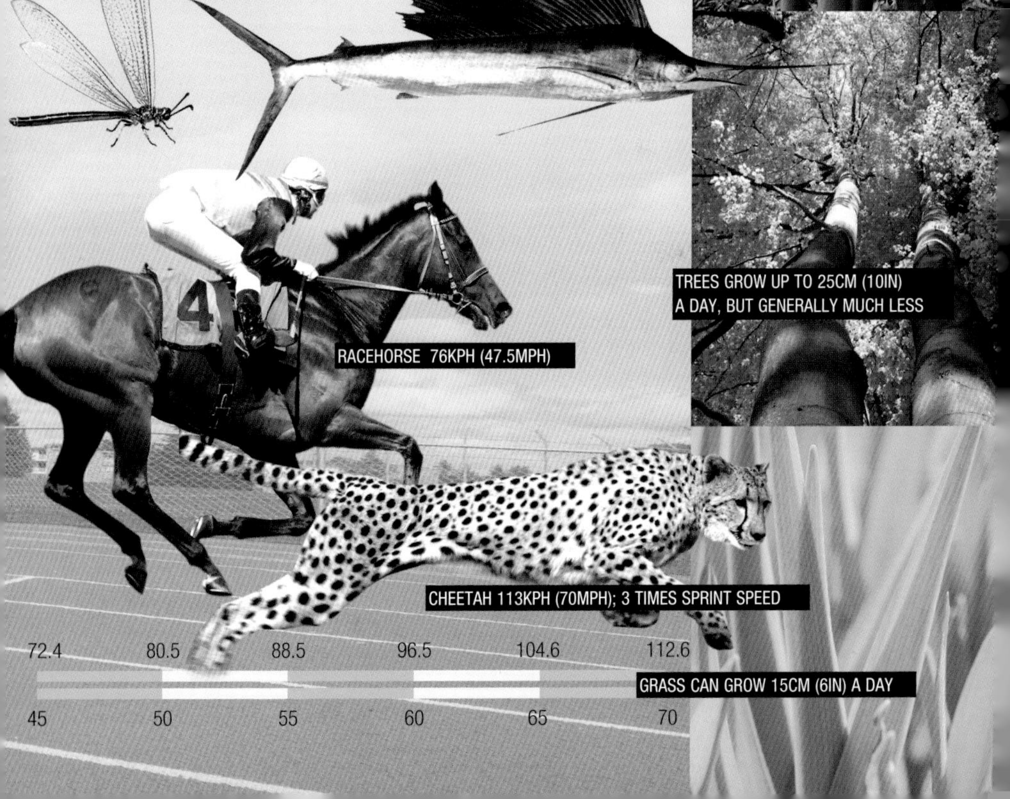

DRAGONFLIES 64KPH (40MPH)

SAILFISH 113KPH (70MPH)

TREES GROW UP TO 25CM (10IN) A DAY, BUT GENERALLY MUCH LESS

RACEHORSE 76KPH (47.5MPH)

CHEETAH 113KPH (70MPH); 3 TIMES SPRINT SPEED

72.4 80.5 88.5 96.5 104.6 112.6

GRASS CAN GROW 15CM (6IN) A DAY

45 50 55 60 65 70

QUITE AN ACHIEVEMENT

There seems to be no limit on the speeds humans can achieve. Records are being broken year after year, some by astonishing margins. When Roger Bannister ran the 4-minute mile in May 1954, it was something of a miracle; these days it's commonplace. We're so blasé about the speeds; it doesn't do any harm to put them into context.

251.4kph (156mph)

SKIING SPEED RECORD

Current world records: For the 100m sprint, runners clock up a top speed of about 42.5kph (26.5mph), enough to pass cars in many urban areas. However, some cyclists, on the flat, have topped 129kph (80mph), faster than the cheetah and well over motorway limits. If that wasn't enough, if they're following a pace vehicle, they can double that speed, and indoors on a roller Bruce Burford broke all records with a 334.6kph (207.9mph) ride – much the same as a high-speed train.

We're still not great in the water though. Our best effort for freestyle over 50m is only half as fast again as a brisk walk, and comes nowhere near the speed of most fishes.

SKYDIVING SPEED RECORD

511.63kph (317.9mph)

CYCLING SPEED RECORD (FLAT SURFACE, UNPACED)

133kph (83mph)

Quick service

Other sports events produce some impressive speeds, too. A professional tennis player's serve is often over 160kph (100mph) and the record, unbroken since 1931, is 263.3kph (163.6mph) – about 1.5 times the speed of the fastest baseball pitcher's fastball.

0.220 SWIFT RIFLE BULLET SPEED
1,963KM/SEC (1,220 MI/SEC)

Faster than a speeding bullet

Not content with the speeds we can manage under own steam, we've invented all kinds of forms of transportation, and they're getting faster at an accelerating rate. Practically all cars on the road are capable of a cheetah's pace (which many countries have adopted as their motorway speed limit), and there's a production car capable of nearly 4 times that speed.

RE-ENTRY SPEED OF SPACE SHUTTLE
16,155 MPH (26,000KPH)

TYPICAL MODERN HIGH-SPEED TRAIN
199MPH (320KPH)

What about the so-called bullet train?

Most modern high-speed trains cruise along at about 320kph (200mph), which although fast, is less than half the speed of the slowest shotgun pellet. However, assuming an average bullet velocity of around 1,600kph (1,000mph), there are plenty of contenders for the honour of matching bullet speed – in the air. Passenger jets are easily capable, but tend to cruise just below it; Concorde regularly more than doubled it. The really fast movers, however, are spacecraft. They have to be, because the escape velocity from Earth is 40,320kph (25,000 mph). They are faster than a speeding bullet, and maybe even faster than Superman himself.

HOT & COLD RUNNING WATER

Rivers run. How fast they run is, like animals running, dependent on their size and the terrain they run across. When they're small, and rushing down steep slopes, they go a lot quicker than when they get down to sea level. Although it may seem to move at quite a pace, even a fast-flowing stream in flood is only going at about the speed of a flying bee (just as well for fish such as salmon, which need to swim upstream to breed); and a typical average speed for rivers is equivalent to a briskish walking pace.

Glaciers, being in effect frozen rivers, move much more slowly, if at all. In fact, their maximum speed is estimated to be around 30m/100ft a day, but that's still about 4 times the speed of a tortoise. Ocean currents aren't exactly hurrying either. Most of the main currents flow at around twice a snail's pace, but the Gulf Stream 'races' across the Atlantic, reaching speeds in excess of 4kph (2.5mph) – a leisurely stroll.

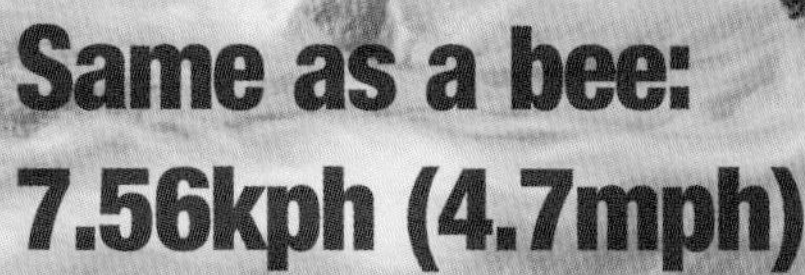

Same as a bee: 7.56kph (4.7mph)

FAST-MOVING NILE RIVER DURING INUNDATION SEASON

SNAIL

Twice as fast as a snail: 0.216kph (0.13mph)

SLOW-MOVING NORTH PACIFIC CURRENT

Moving mountains

It's not just water that travels across the surface of the earth. The continents are on the move too, albeit very slowly. Continental drift, the movement of the tectonic plates, varies from place to place up to about 10cm (4in) a year, but averages about 5cm (2in) a year. That may be imperceptibly slow – even a snail moves a hundred times faster – but over a period of decades or centuries it's enough to have an impact. Ask anyone who lives on a fault line.

An ill wind

There are a number of words to describe wind speeds: zephyr, breeze, gust, gale, hurricane – the list goes on. They give an idea of the force of the wind, but it took Rear-Admiral Sir Francis Beaufort (1774–1857) to come up with a scale that pinned it all down and describe the effects of the wind as well as its speed. It's even more illuminating if you compare it with other natural speeds.

IT MAKES YOUR HEAD SPIN

Something to bear in mind when you're talking about speed is that it's always relative to something. You can get some idea of what that means by thinking of a passenger on a train travelling at 80.4kph (50mph). If he gets up and walks to the front of the train, how fast is he walking? Right – his speed relative to the train may be 4.8kph (3mph), but relative to the track, it's 85kph (53mph). And if he walks to the back, it's 76kph (47mph).

1,670kph/1,038mph

The rotational speed of the earth at the equator, which decreases in mid-latitudes

It's all relative: Most of the time, we mean speed relative to the ground, which, for most practical purposes, is reasonable, but don't forget the earth is moving, too. It's rotating once a day, which means that at the equator you'd be hurtling around at almost twice the speed of a jet plane. At the poles, you'd be stationary, doing a very slow pirouette.

The earth is also travelling through space as it makes its way around the sun. It takes a year to complete one orbit, maintaining an average speed of 106,768.8kph (66,343.056mph) or 29.658 kilometres per second (18.429 miles per second) relative to the sun. Which is itself orbiting the galaxy, in a universe that is expanding, at heaven knows what speed …

As fast as can be: Nothing travels faster than light. You'll just have to take Einstein's word for that. Light travels at an unimaginable speed – 299,792,458 metres per second (983,571,058 ft/sec) or 300,000 kilometres per second (186,000mi/sec), more than a million times the cruising speed of a passenger jet.

Light travels from the sun to Earth in 8 minutes, 18 seconds 186,000km/sec (186,000mi/sec)

Is there anybody out there?

News travels fast, they say. However, not even radio waves can travel faster than light. So if there is intelligent life somewhere out in the cosmos, it might take a long time for us to hear about it. Our nearest neighbouring star, Proxima Centauri, is 4.2 light years (L.Y.) away, and there's no sign of a habitable planet there. The nearest possible candidate is more than 10.5L.Y. distant, so communication would be slow: 21 years before you got a reply to your message. The closest planets with intelligent life (as we know it) may be hundreds, or even thousands of light years away. If we got news from them, it would be centuries out of date; and who knows what state our planet will be in by the time they get any message from us.

THE SOUND BARRIER

When Concorde, the first supersonic commercial jet plane appeared, everyone agreed it was an engineering marvel. Faster than the speed of sound. Fantastic! However, few of its fans would admit to not knowing what that really meant.

One of the problems is that the speed of sound is not constant: it is different through air, water and other fluids, and also depends on temperature. For our rough comparison purposes, we can put it at about 1,240kph (770mph), on a fine, dry day. That's 11 times the speed of a running cheetah, or about 1.4 times the cruising speed of most passenger airliners.

However, things get a bit strange when we get under water. Where everything else tends to get slower, sound actually travels faster. About 4.5 times faster. The warmer the water, the faster it is – and it travels even quicker in salt water than fresh.

How fast is sound in space then?

Trick question. Because space is a vacuum, sound can't be transmitted across it, so its speed is zero. It's true: if you scream, no one will hear you.

Like greased lightning

Because the speed of light is so incredibly fast, it's near enough instantaneous over short distances. In a storm, you see the flash of lightning practically the moment it happens. However, sound travels relatively slowly through the air (nearly a million times slower than light), and the thunderclap doesn't reach your ears until a bit later – about 3 seconds for every kilometre (or 5 seconds later for every mile) you are away from the storm.

9 TEMPERATURE

Blowing hot and cold: Because we have a very limited tolerance of hot and cold, we can only survive extremes of temperature relatively close to our body temperature – our vocabulary for describing temperatures is limited, too. There's warm, mild, hot, cold, boiling, freezing and not much else. And once we go outside our comfort zone, it's either just very hot, or very cold.

To get a little more precision into our descriptions, we need some comparisons with things we understand – even if we couldn't survive them – like cooking and deep freezing. From there we can get a handle on some of the highest and lowest temperatures in the world around us, and beyond.

SOME LIKE IT HOT

Like the little bear's porridge, the earth is not too hot, not too cold, but just right for the life forms that inhabit it. Or maybe we're just right for the planet, because it's what we evolved in. Whichever way you look at it, there's a fairly limited range of temperatures in the world (on the surface, at least: see pages 102–103), even though the extremes might seem, well, extreme by most people's standards.

Dallol, Ethiopia 34.4°C (94°F)

There's room for some vastly different climates on the earth. The average temperature of the world is a pleasant 15-20°C (60–70°F), with different climate zones ranging from the icily polar, averaging as low as -68°C (-90°F) on the Antarctic Plateau, to a torrid average of 34.4°C (94°F) in Dallol, Ethiopia – with plenty of temperate climates in between.

Antarctic plateau -68°C (-90°F)

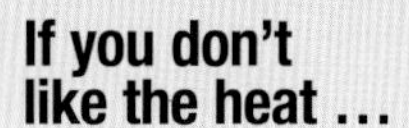

If you don't like the heat …

The world's hottest recorded temperature, at Al 'Aziziyah in Libya in 1922, was a sizzling 57.8°C (136°F), although there have been claims of as high as 70.7°C (160°F) in the Lut Desert in Iran.

Libya

34.4°C (136°F)

At the other end of the scale, the coldest recorded temperature was, as you might expect, deep in the Antarctic: -89.4°C (-129°F) – about 130°F (75°C) lower than a domestic freezer. The total range between these two extremes is 147°C (265°F).

Antarctic

-89.4°C (-129°F)

AS HOT AS HELL

It's easy to forget that the earth is really a lump of molten rock, just beginning to cool down. Unless you live near a volcano, of course, where some of the molten rock squirts through the planet's thin crust. This molten lava comes to the surface at a temperature of about 1,700°C (3,000°F), about the same as a modern blast furnace, and 10 times the heat you would roast your Sunday dinner in.

However, the lava cools down a little on its way out. It comes from the mantle, the layer of the earth beneath the crust, where the average temperature is half as much again. Beneath that is the core, and that's where it gets really hot: the outer core is about 4,982°C (9,000°F) – almost the same as the surface of the sun, and the inner core is probably even hotter. You can see where Dante got the idea for his Inferno.

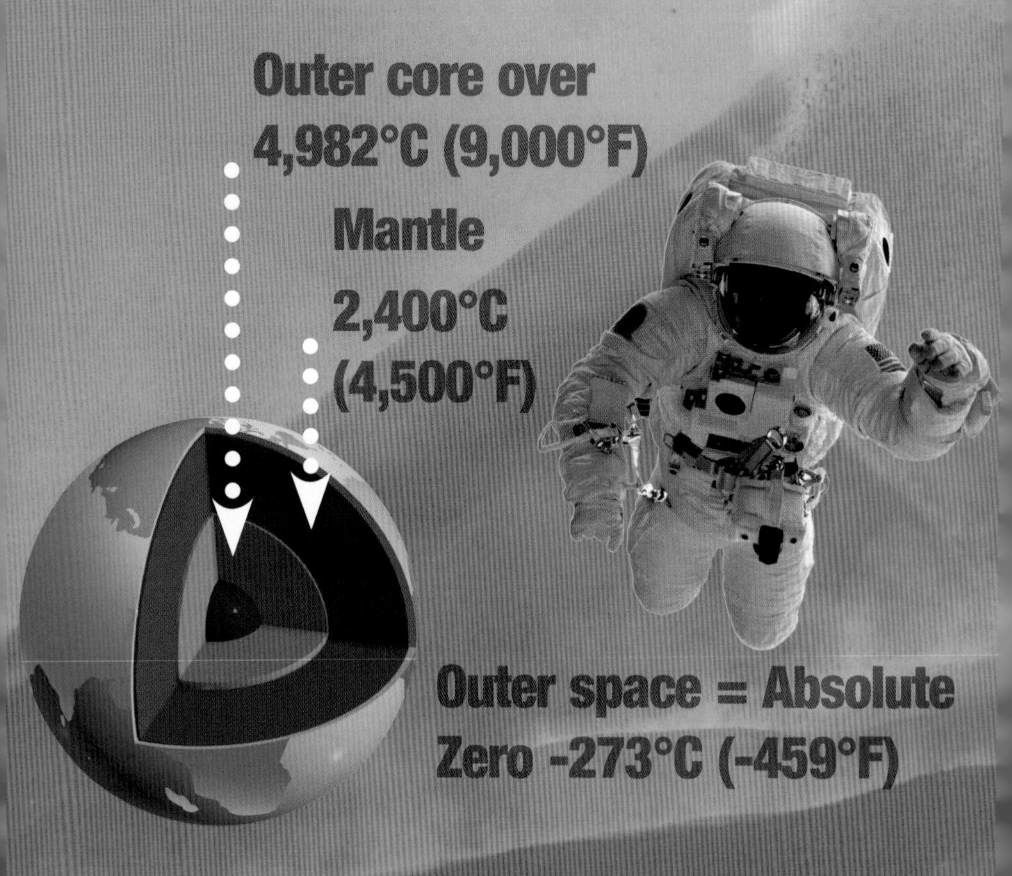

Rising temperatures

Temperature drops to nearly absolute zero once you get into outer space. So it kind of makes sense to think that the higher you go (that is, farther from the earth's surface), the colder it gets. Which is true, up to a point … but some strange things happen along the way.

When you're in a jet travelling to your holiday destination, little do you know that it's cold outside. Really cold. After an airliner takes off and makes it way through the troposphere, the outside temperature gradually drops, to lower than Antarctic levels of -68°C (-90°F) when it reaches cruising altitude just at the bottom of the stratosphere. However, if you continued to rise, it would get warmer again (relatively, anyway) reaching around freezing at the base of the mesosphere. Above that, the temperature plummets down to around -90°C (-120°F) as you ascend through to the thermosphere, then suddenly rises rapidly to anything between 330–1,500°C (630–2,700°F) at the thermopause, where the thermosphere meets the exosphere. Beyond that, temperature fluctuates wildly between night and day – from almost absolute zero to many thousands of degrees.

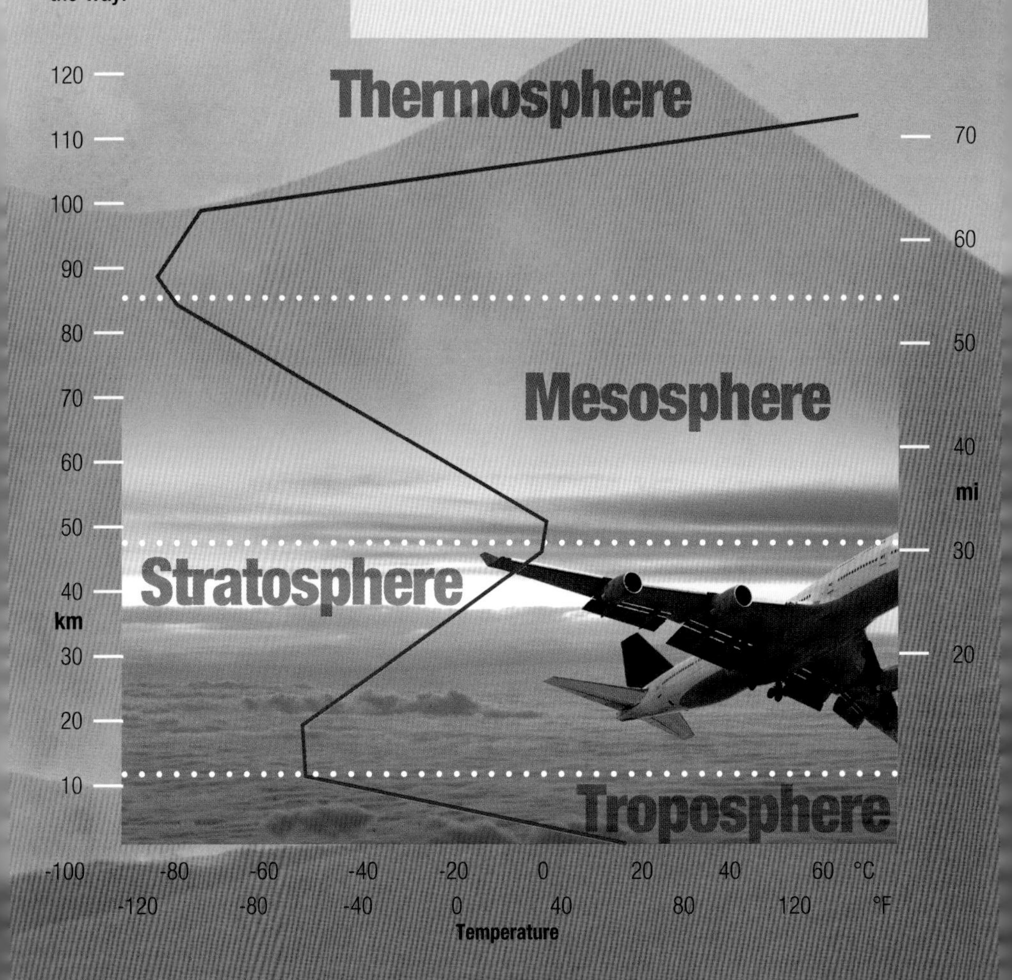

BODY HEAT

We're hot-blooded animals, we humans. In fact, to be more accurate, we're warm-blooded. Like all mammals and birds, our bodies maintain a pretty constant temperature; just a couple of degrees above it and we're complaining of a raging fever. Yet, hibernating mammals can lower their body temperatures to levels that would be considered hypothermia in humans. In general, our body's natural thermostats keep our temperatures within limits suited to the temperate climates we live in.

Body heat

Reptiles, amphibians and fish have to survive harsher conditions. Although they're called cold-blooded, their body temperatures cover a wide range – they have no internal heating or cooling devices, so their body temperatures are very much dependent on their environment. This arrangement means that some fish can swim in ice-cold water, and many reptiles can tolerate the hot desert sun.

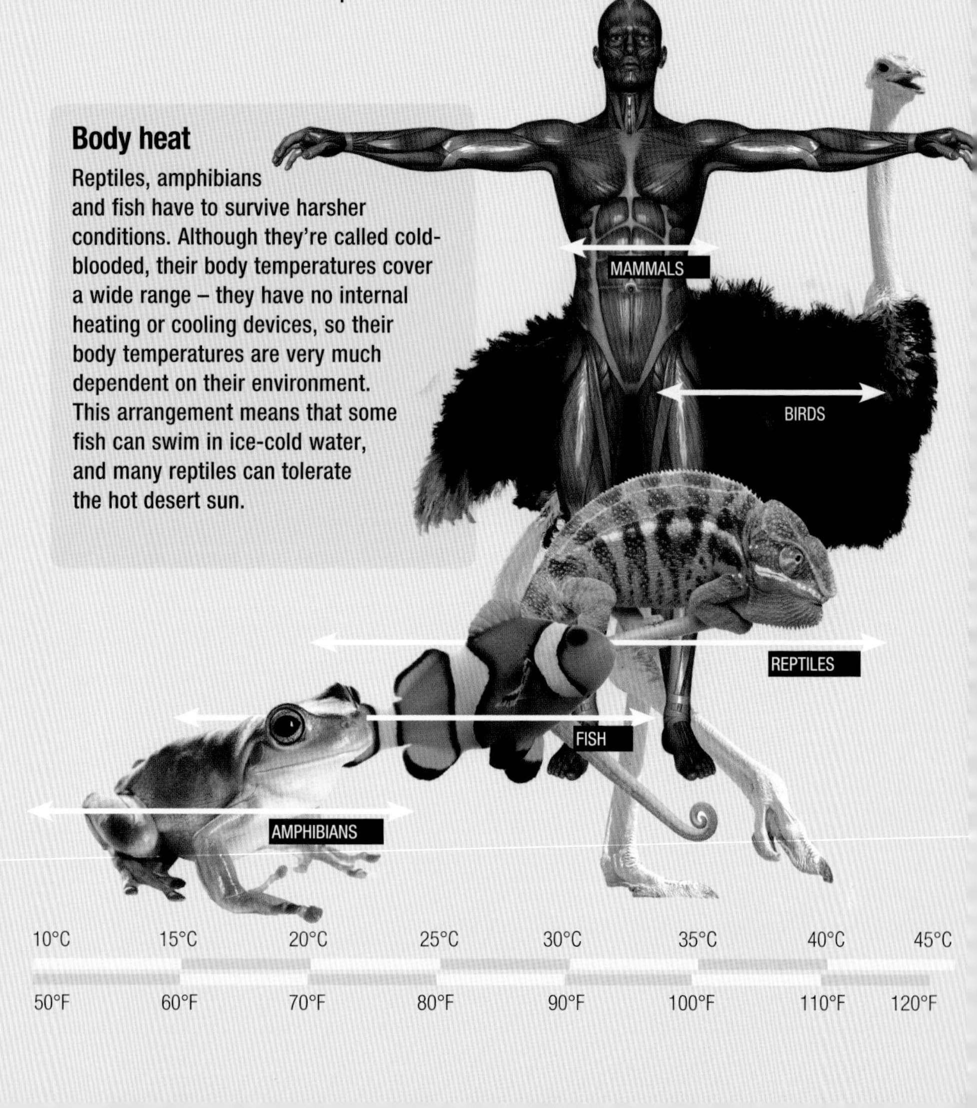

A climate to die for

There seems to be an innate urge for humans to explore. Having discovered a good part of our own planet, and having found good reasons why we don't live in some parts of it, we've expanded our horizons and started exploring other worlds, perhaps in the hope we might find somewhere else to settle.

What are the chances? Pretty slim, if the planets in the solar system are anything to go by. Leaving aside the small matter of a lack of breathable atmosphere, the temperatures alone make most of them uninhabitable for any kind of life we know. Our nearest neighbour, Mars, seems the most hospitable climate-wise: nice warm summer days up to body temperature, but winter nights colder than anywhere on Earth. Mercury? Ranging from heat enough to melt lead down to a low that would liquefy oxygen, this doesn't seem a good option. As for the more distant planets, forget it – unless you're interested in cryogenics.

On Neptune, nitrogen is considered a solid, not a gas

FREEZING COLD, BOILING HOT

We're all familiar with the freezing (or melting, depends if you're cooling or heating it) point of water, and its boiling point, too – we even probably know what temperatures they are. What we may not realise is that those are the points where a substance (in this case water) changes from one state to another – from solid to liquid, and from liquid to gas: from ice to water, from water to vapour.

It's difficult to believe that everything has freezing/melting and boiling thresholds and can exist in those three states – even stuff we consider to be really solid such as rock, and stuff we think of as gas including hydrogen, because we generally see them only at everyday temperatures. However, have a look at mercury, and it's easier to believe that this is a molten metal, which won't solidify (freeze) until it gets down to -39°C (-38°F).

Freezing/melting and boiling points

	Freezing point °C (°F)	Boiling point °C (°F)
Hydrogen	-259 (-432)	-253 (-423)
Oxygen	-219 (-362)	-183 (-297)
Chlorine	-101 (-150)	-35 (31)
Mercury	-39 (-38)	357 (675)
Water	0 (32)	100 (212)
Phosphorus	44 (112)	280 (536)
Sulphur	113 (235)	445 (833)
Lead	328 (622)	1,740 (3,164)
Tin	232 (450)	2,270 (4,118)
Silver	962 (1,763)	2,210 (4,010)
Copper	1,083 (1,981)	2,566 (4,651)
Gold	1,064 (1,947)	2,900 (5,252)
Iron	1,536 (2.796)	3,000 (5,432)
Carbon	3,550 (6,422)	4,825 (8,717)
Tungsten	3,410 (6,170)	5,660 (10,220)

10 ENERGY & POWER

E = mc²: Einstein's famous equation finally solved the mystery of energy. Or at least for the scientists it did. For the rest of us, it's still very much a grey area. We use energy of one kind or another every day, but we've no idea what it actually is.

Perhaps that doesn't matter, as long as we have some way of measuring it. We know roughly how far a tankful of petrol will take us, or how much (in monetary terms) electricity we've used in a month. However, when we start thinking about the pulling power of machines or the energy consumption of nations, we're often in the dark. Putting them in the context of our everyday use of energy, however, may shed a little light on the matter.

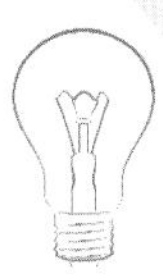

HORSE POWER

James Watt got there before us. He needed a unit to describe energy, and came up with the horsepower (hp). It really can't be improved – it's easily visualised, and sort of means something. More precisely, it means the energy used to lift a weight of 550lb 1 foot in 1 second. A horse can do this, apparently (in fact, horses are capable of doing more when they put their minds to it – they have a capacity of about 15hp).

The metric equivalent: This is the energy needed to raise 75kg 1 metre in 1 second, about 745.7 watts. It comes to much the same thing, for our rough comparison purposes.

So, when you buy a 100hp tractor, you know it has the power to lift 55,000lb 1 foot in 1 second. More interestingly, we humans can, in short bursts, produce as much as 2.5hp; an 8-man tug-of-war team could manage about 20hp. No match for the tractor, but enough to beat even the strongest horse.

A third horse

Not content with the units available, some witty engineers added another, with a value of 250 watts. At about one-third of a horsepower, it's known as the donkey power.

3 donkeys = 1 horse

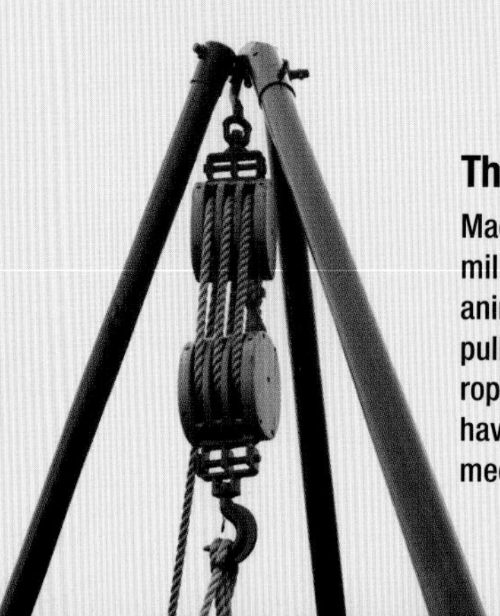

The advantage of machines

Machines to help pull and lift heavy weights have been used for millennia. Before the invention of engines and motors, human or animal power was converted into a more useful force by simple pulleys and complex block-and-tackle devices: by pulling on a rope and a system of pulleys, it's easier to move a load, but you have to pull it correspondingly farther, in a ratio known as the mechanical advantage.

Hitching a lift

We've come a long way since Watt invented the horsepower. There are all kinds of machines for lifting, pulling and pushing loads, many of them beyond even his wildest dreams. Walk around most warehouses and you'll see forklift trucks capable of lifting anything up to about 5 tons, and outside you might see one lifting a loaded container of up to 50 tons. Or that job might be left to a crane.

Human ingenuity has provided us with a huge variety of cranes, for loading ships, lifting building materials and even putting whole sections of bridges into place – the largest of them can lift several thousand tons.

For pulling or pushing, there are tractors. Little ones, of about 20hp, through to the 100hp model you'll find on a farm, to heavy industrial beasts of nearly 600hp. However, these are as nothing compared to the little tugboats nudging ocean liners and tankers into place: harbour tugs have 700 to 3,500hp engines, and their deep-sea bigger brothers can produce more than 25,000hp.

THE POWERS THAT BE

Energy doesn't come from nowhere. For small amounts, we can use the chemical energy supplied by batteries, but to provide the power to run our homes and industries, we need some kind of fuel. For transportation, that's still almost exclusively oil, despite increasingly dwindling supplies. For the rest, there are power stations that convert the energy of various fuels into electricity.

To meet demand, the output of power stations has to be in megawatts (M.W., millions of watts) or gigawatts (G.W., billions of watts), so efficiency is vital. Different fuels vary in their efficiency – coal-fired power stations are relatively inefficient, natural gas power relatively efficient. The jury's still out on the efficiency and economic viability of nuclear power, quite apart from safety issues.

Alternatives

Worries about dwindling supplies of fossil fuels, pollution and global warming have prompted renewed interest in alternative sources of energy, especially clean, renewable ways of generating electricity. Until recently, the arguments against alternatives have been financial, while fossil fuels were cheap, but they are now being investigated more seriously out of necessity.

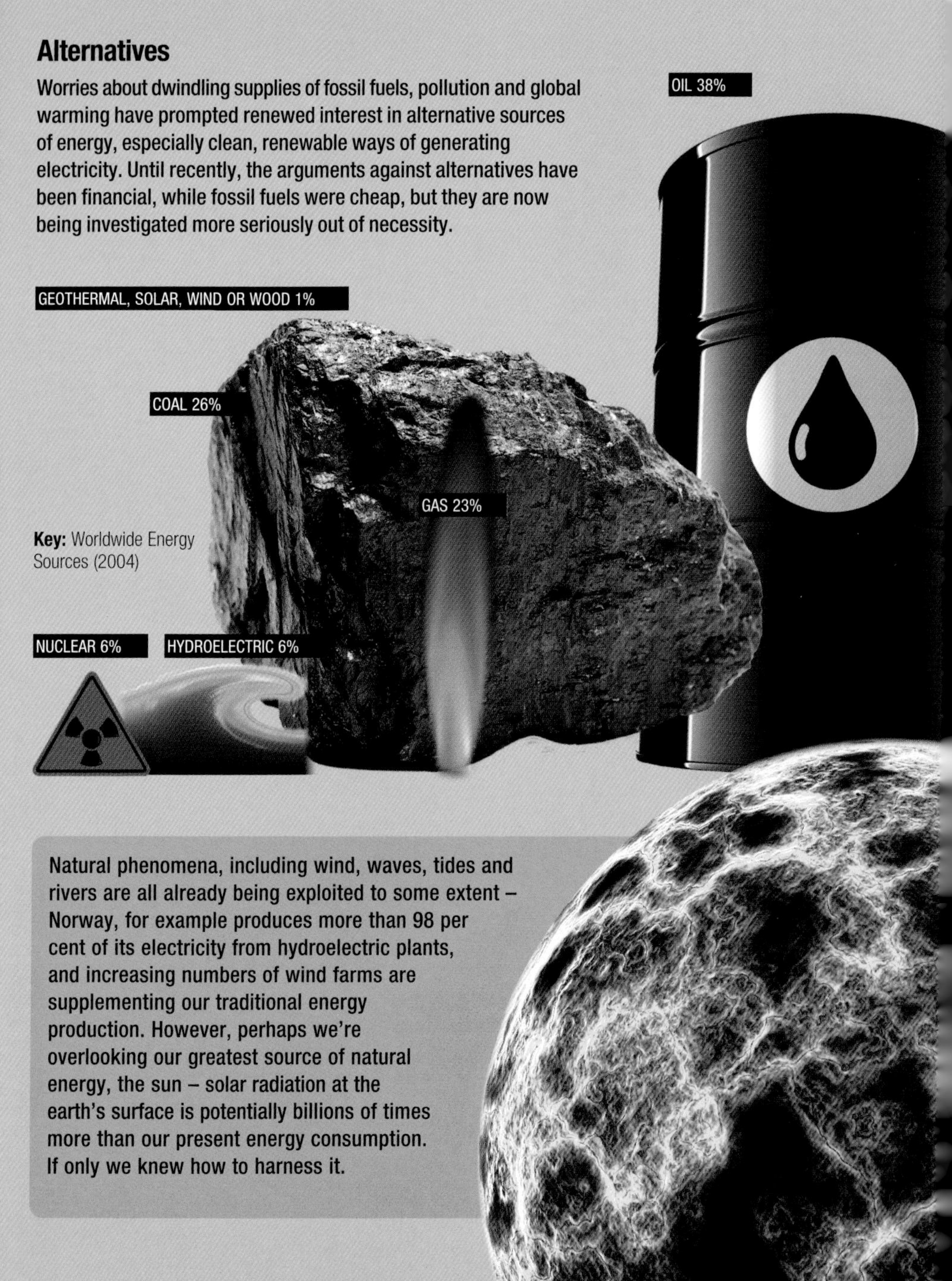

Natural phenomena, including wind, waves, tides and rivers are all already being exploited to some extent – Norway, for example produces more than 98 per cent of its electricity from hydroelectric plants, and increasing numbers of wind farms are supplementing our traditional energy production. However, perhaps we're overlooking our greatest source of natural energy, the sun – solar radiation at the earth's surface is potentially billions of times more than our present energy consumption. If only we knew how to harness it.

WHAT WATT?

James Watt gave us a useful reference point in the horsepower, but it's not so good at measuring our everyday energy use. Nevertheless, he still gets to be immortalised in the watt (W), which most countries use to measure their electricity production and consumption.

USA 1,0381.2W

UK 5218.2W

Total Energy Consumption per capita per annum:

1 light bulb = 100 watts

1 oven = 1,000 watts

What is a watt?

To get an idea, think of a light bulb – one of the old ones, before we started using the energy-saving type. Chances are, the main light in your living room would have a 100-watt light bulb, which (unsurprisingly) uses 100 watts of power. Now, a coffee-maker uses about 10 times the power (1,000 watts, or 1 kW). If you know the wattage of an appliance, you can work out which ones are causing the big bills each month.

Using the wattage of domestic appliances puts our consumption into some kind of perspective. Total energy consumption per capita in the USA, for example, is around 10,000W; on that basis, a four-person family would be using the equivalent of 400 light bulbs, night and day.

Bangladesh 214.4W

Ethiopia 370W

Singapore 6,870.6W

Consumer society

The threat of global warming has made us more conscious of our energy consumption and the effects of our collective carbon footprints. It wouldn't do any harm to look at it in a wider context, too; measuring it against production, and comparing that with other countries around the world.

In the USA, for example, each person is using energy at the rate of over 1,000 watts; almost twice what many Europeans use, and 50 times the consumption of the average Bangladeshi. Worse still, the USA uses almost 1.5 times the energy it produces. Middle Eastern oil-producing countries produce as much as 5 times what they consume, despite being high on the list of per capita users of energy.

As the oil runs out, we face some tricky decisions: do we decrease our demand for energy, or find new ways of producing it? And are there any clean, sustainable options?

FOOD FOR THOUGHT

Our bodies use energy too, even when we're sleeping. We get the energy from the food we eat, not just for activities like running and physical work, but also to keep our vital organs going and maintain our body temperature (that's why cold-blooded animals eat less than we do). The energy in foodstuffs is measured in Calories (not to be confused with calories, with a small c, which are 1,000 times smaller), as anybody on a reducing diet will tell you.

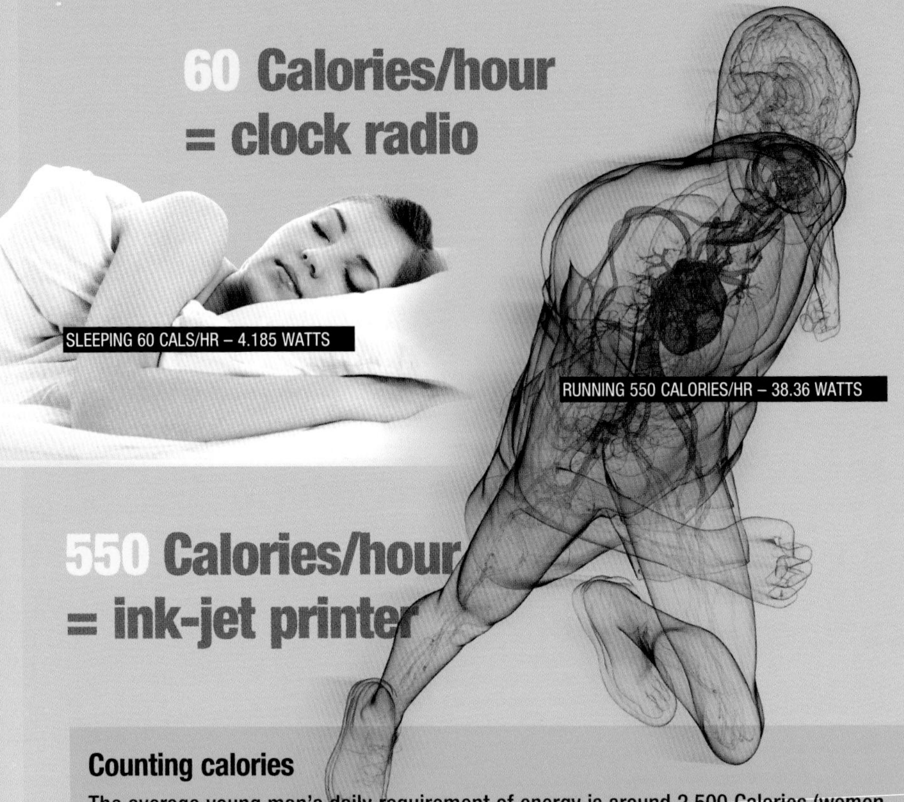

Counting calories

The average young man's daily requirement of energy is around 2,500 Calories (women need a bit less, about 2,000 Calories); in energy terms, that's much the same as is needed to run a 7-watt energy-saving light bulb for 24 hours. To get that, he or she can choose between a whole variety of different foods, all with different Calorie values; from butter (200g/8oz would give around 1,750 Calories) through lean meat (about 550 Calories in a 450g/8oz steak) to vegetables (30 Calories to a 200g/8oz serving of boiled cabbage).

Spending power

We all need food. And the energy it gives can be measured. So, it's a constant – unlike money, the value of which fluctuates according to the whims of market economies. As a result, some economists use food, or at least a well-known variety of food, as a measure of economic spending power, thus avoiding the complications of inflation of the pound, yen or euro. Their benchmark for measuring the purchasing power parity of different countries is often the Big Mac, the cost of a McDonald's Big Mac burger, by which prices of other commodities and incomes can be measured.

Other units such as the loaf of bread are sometimes used, especially when comparing prices or incomes over a period of time; economists in the UK prefer to use the Mars Bar as their reference point, while others have suggested the iPod, or Ikea's Billy bookshelf.

THE BIG BANGS

As with so many things, man is a long way behind nature in terms of energy, especially when it comes to destructive power. We've developed frighteningly powerful 'weapons of mass destruction' such as the hydrogen bomb, but frankly they're mere squibs compared to what nature has to offer.

Our best effort has produced the so-called Van hydrogen bomb, which unleashed 50 megatons of energy (big bangs like this are measured in TNT equivalent – how much TNT would cause the same explosion) when the Russians tested in 1961. Three-quarters of a century earlier, the volcano in Krakatoa erupted with a force 3 times that of the Russian h-bomb, and it's estimated that the Tambora eruption in 1815 was more than 100 times more powerful.

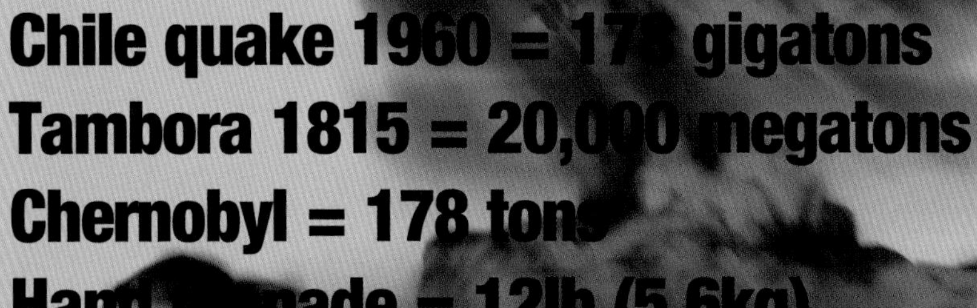

Chile quake 1960 = 178 gigatons
Tambora 1815 = 20,000 megatons
Chernobyl = 178 tons
Hand grenade = 12lb (5.6kg)

Earthquakes put our destructive attempts into perspective, too. The Chernobyl disaster in 1986 would have only merited a 3.5 on the Richter scale ('minor'), and even a 1 kiloton A-bomb would only register a 4 ('light'). The earthquake in Valdivia, Chile, in 1960 was the largest ever recorded, with a magnitude of 9.5 on the Richter scale; the equivalent of a 178 gigaton bomb, or 3,560 times the power of the biggest bang we've managed. So far.

SOUND

Have you heard this one? Quantifying sound is a problematic business. One man's meat is another's poison, and music to some is noise to others. And what do phrases such as 'twice as loud' really mean? Or 'twice as high'? It's all a question of perception. There are particular ways of measuring the intensity of sound, using a scale of decibels (dB), and the frequency in hertz (Hz) – but they're just as abstract to most of us as the sound they're describing.

It may help when you make an official complaint about the party next door if you can tell the authorities that the noise reached 120dB, but when you tell your mates, it will mean more if you say it was as loud as a jet preparing for take-off.

LOUD & CLEAR

Scientists and engineers can measure sound levels incredibly accurately, and give their results in decibels. However, as with so many other scientific measurements, this term is practically meaningless to the uninitiated. To make sense of it, we need some examples for comparison.

The decibel scale starts (sensibly) at 0dB, which is the threshold of audibility – in other words, the very quietest that a human can hear – and goes up to about 140dB, the loudest sounds we're likely to come across. In between, of course, there's a whole range of loudness. For the non-scientific, the scale goes from silence to a whisper, a conversation and a shout, to a chain saw, thunderclap and gunshot.

Coincidentally, each 10dB step up the scale corresponds to a perceived doubling of the loudness of sound (although that's a bit like saying something's twice as red) – so a 100dB sound appears 'twice as loud' as one of 90dB.

Chain saw
100dB

Whisper
30dB

Shout
80dB

Rustle of leaves
20dB

As sound as a bel

The decibel scale is useful for making comparisons, but beware! It's a logarithmic scale. Now, without getting too technical, let's see what that means …

Take a barely audible sound such as a very quiet breath at 10dB. To increase the loudness to 20dB (equivalent to the rustle of leaves) you'd need 10 people breathing quietly. To get to the level of a whisper, 30dB, you'd need 100 quiet breathers (10 times again). Get the idea?

Similarly, one person shouting gives us 80dB, so 10 shouters would make a 90dB noise, 100 would get us up to 100dB and a crowd of a thousand would be cheering at 110dB. So you'd need more than 10,000 to be heard over a rock band in the middle of a performance.

76 trombones

So, exactly how loud are 76 trombones? The answer might surprise you. A single trombonist can play (remarkably easily, unfortunately) a volume of about 90dB. You'd think two of them would play twice as loud – but you'd be wrong. To double the perceived loudness of the sound (to 100dB, remember?) there would have to be 10 times the number of trombones, and to get up to 110dB you have to multiply that by 10 again – that would be 100 trombones. So 76 trombones playing at 90dB would make a sound of about 108dB, just a bit quieter than a thunderclap (110dB), and enough to cause some hearing loss.

THE HIGH Cs & THE LOW Cs

The limits of human hearing are not just determined by loudness; we can't hear very high or very low sounds, no matter how loud they are. We call the sounds above and below our hearing range 'ultrasonic' and'infrasonic', as if sounds outside our hearing range didn't exist – but, of course, they do, and other animals can often hear them.

Think of the 'silent' dog whistle. We can't hear it because the sound is so high, but it's within a dog's hearing range. Bats and dolphins also have acute hearing at the upper end of the sound spectrum, picking up frequencies way above our hearing range (about three octaves above, and five octaves above the top notes of a piano). This, coupled with the ability to emit ultrasonic noises, gives them the ability to 'see' by echolocation at night or in murky waters.

At the other end of the scale: Elephants and whales communicate using very low frequency sounds, which can carry over very long distances, and very few other animals can hear them. Certainly not most birds, anyway, which have surprisingly limited aural capabilities – their hearing is pretty much limited to the frequency range of their own song.

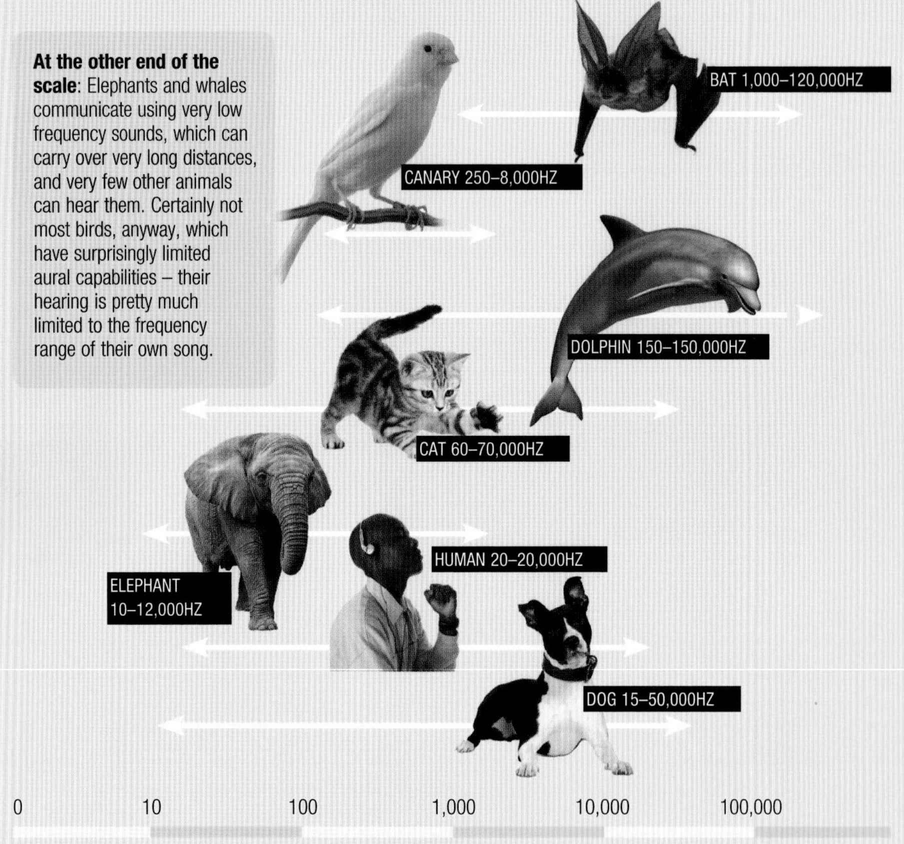

Note	Frequency
A0	27.5
B0	30.868
C1	32.703
D1	36.708
E1	41.203
F1	43.654
G1	48.999
A1	55.0
B1	61.735
C2	65.406
D2	73.416
E2	82.407
F2	87.307
G2	97.999
A2	110.0
B2	123.47
C3	130.81
D3	146.83
E3	164.81
F3	174.61
G3	196.0
A3	220.0
B3	246.95
C4	261.63
D4	293.66
E4	329.63
F4	349.23
G4	392.0
A4	440.0
B4	493.88
C5	523.25
D5	587.33
E5	659.25
F5	698.46
G5	783.99
A5	880.0
B5	987.77
C6	1,046.5
D6	1,174.7
E6	1,318.5
F6	1,396.9
G6	1,568.0
A6	1,760.0
B6	1,979.5
C7	2,093.0
D7	2,349.3
E7	2,637.0
F7	2,793.8
G7	3,136.0
A7	3,520.0
B7	3,951.1
C8	4,186.0

Music to my ears

Although our hearing range doesn't match up to bats and dolphins, it's more than adequate, and it gives plenty of scope for listening to music. Each of the singing voices (soprano, alto, tenor, bass) covers a range of roughly two octaves; but as if that wasn't enough, we've made musical instruments that produce sounds from the very lowest (the contrabassoon) to the highest (the violin) limits of audibility.

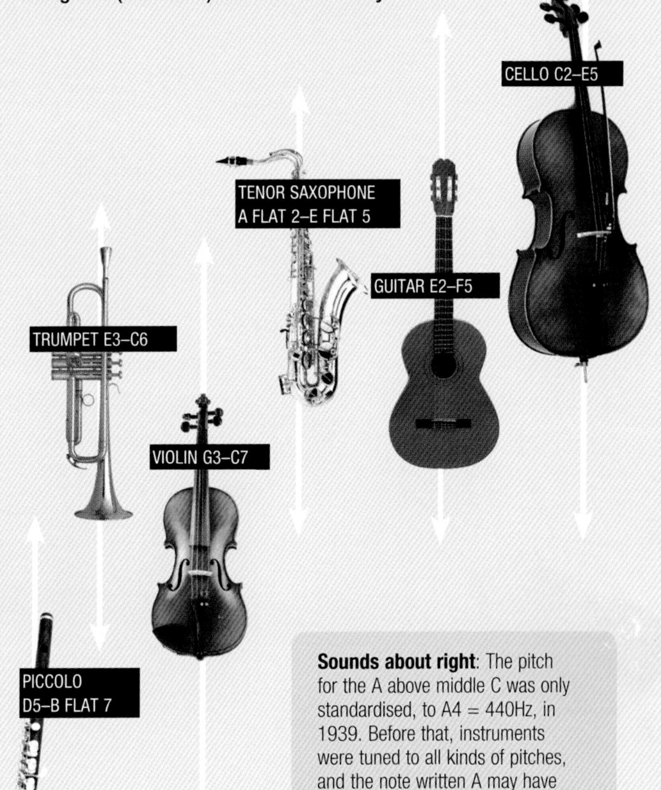

Sounds about right: The pitch for the A above middle C was only standardised, to A4 = 440Hz, in 1939. Before that, instruments were tuned to all kinds of pitches, and the note written A may have been as much as four semitones (4 consecutive notes on the piano) different from today's A. If you play an A on a very old organ for instance, depending on when and where it was made, it could sound as anything from 370Hz to 466Hz.

TABLE OF EQUIVALENTS

Note: These are units that have been mentioned in the book, plus a few extras that could be useful and might encourage you to find some meaningful ones of your own. Most are approximate measures, and equivalents in SI and U.S. Customary units are merely guideline figures. Calculations using these units are often rounded up or down, sometimes quite dramatically, for convenience sake. We're talking rough and ready comparison, remember?

LENGTH & DISTANCE			
Handspan	span	23cm	9in
Pace	pace	0.9m	3ft
Leonardo	Len	1.83m	6ft
Parked car	Pcar	5m	15ft
Bowling alley	alley, al	20m	66ft
London bus length	bus long, lbl	9m	30ft
Five-minute walk	F.M.W.	500m	1,500ft
Marathon	Mar	42km	26mi
Nile	nile	6,650km	4,132mi
Lunar distance	LD	384,403km	238,857mi
Astronomical unit	AU	149,000,000km	92,580,000mi

AREA			
Pinhead	PinH	1mm²	0.00155in²
Shirt button	button	100mm²	0.155in²
CD or DVD		113.1cm²	17.5in²
Dinner plate	plate	500cm²	78in²
A4 paper		630cm²	97.7in²
Double bed	bed	2.5m²	27ft²
Parking space	P-space	10m²	108ft²
Boxing ring	B.R.	40m²	430ft²
[Football field/pitch*]			
Wales	Wal	20,000km²	8,000mi²
Wyoming	Wyo	250,000km²	100,000mi²
Mediterranean Sea**	Med	250,000 km²	100,000mi²

* Depends whether you're a football fan or American football fan

** This is almost identical to the Wyoming, but provides an alternative for marine areas

HEIGHT & DEPTH

Monsieur Eiffel	MEiff	1.83m	6ft
Story	sto	3m	9ft
Giraffe	gir	6m	18ft
Eiffel Tower	Eiffel	300m	1,000ft
Burj Khalifa	B.K.	828m	2,717ft
Everest	Ev	8,848m	29,030ft

WEIGHT & MASS

Grain of sand	Sgrain, sg	0.01g	0.00035oz
Flea	fl	0.1g	0.0035oz
Tomcat	Tom, T.C.	4kg	9lb
John Doe	J.D.	80kg	175lb
Family car	carweight, fcw	2,000kg	4,500lb
Elephant	El	6 tons	6 tons
London bus weight	busweight, lbw	10 tons	10 tons
Jumbo jet	Jumbo*, Jj	360 tons	800,000lb
Blue whale	Bwhale, bw	150 tons	150 tons

* Not to be confused with an elephant nor, indeed, with an airbus (a unit of population)

VOLUME & STORAGE CAPACITY

Wine glass	Wglass	125ml	4fl oz
Cup	cup	250ml	8fl oz
Wine bottle	Wbot	750ml	25fl oz
Bathtub	Btub	500l	110 gallons
Olympic swimming pool	Opool	2,500,000L (2,500m^3)	88,300ft^3
Phone box	Pbox	2m^3	70ft^3
House	Ho	1,000m^3	35,000ft^3
Container	Cont	43m^3	15,200ft^3
Gas tank	Gtank, Gt	57l	12.5 gallons
Barrel	bbl*	160l	35 gallons
Sydney Harbour	sydharb	0.5km^3	400,000 acre-feet

* Why the double b? Not my idea: this is an already established abbreviation for barrel.

POPULATION		
(Greyhound) busload	bus	50 people
Airbus 380	Airbus, Abus	500
Albert Hall	Albert	5,000
Hollywood Bowl	Hbowl, Hb	15,000
Village	Vill	1,000
Town		100,000
City (Odessa)	Od	1,000,000
Beijing	Beij, Bj	10,000,000
Burma [Myanmar]	Bma	50,000,000
Mexico	Mex	100,000,000
India	Ind	1,000,000,000
World population	Wpop, wp	6,800,000,000

TIME		
Flash	Fl	0.001 second
Second	sec, s	(1 second)
Hour	hr, h	60 seconds
Day	day	24 hours
Year	yr, annum, a	365.242 days
Generation	gen	25 years
Lifetime	life	75 years
Century	C	100 years
Pluto year	Py	250 years
Millennium	ka	1,000 years
Megaannum	Ma	1,000,000 years
Galactic Year	G.Y.	250,000,000 years

TEMPERATURE			
Absolute zero		-273.16°C	-459.67°F
Freezing (water)		0°C	32°F
Body temperature (human)		37°C	98.6°F
Boiling (water)		100°C	212°F
Molten lava		1,730°C	3,146°F
Surface of the sun		5,330°C	9,626°F

SPEED			
Snail's pace	Snp	0.08kph	½mph
Brisk walk	Bw	6kph	3¾mph
Sprint	spr	36kph	22.5mph
Racehorse	Rhorse,	76kph	47.5mph
Cheetah		113kph	70mph
Skydiver (average terminal velocity, belly to earth)		193kph	120mph
Passenger jet		900kph	550mph
Bullet		1,600kph	1,000mph
Speed of sound		1,240kph	770mph
Speed of light		300,000km/s 1	86,000mi/sec

ENERGY		
Donkey power	donk	250W
Horsepower	hp	745.7W
Tug-of-war team	tow	20hp
Tractor	trac	100hp
100-watt light bulb	bulb	100W
2-bar fire		2,000W (2kW)

SOUND	
Breath	10dB
Rustle of leaves	20dB
Whisper	30dB
Library	40dB
Quiet conversation	50dB
Muzak	60dB
Vacuum cleaner	70dB
Shout	80dB
Diesel truck	90dB
Chain saw	100dB
Thunderclap	110dB
Rock concert	120dB
Jet take-off (100m/330ft)	130dB
Gunshot	140dB

INDEX

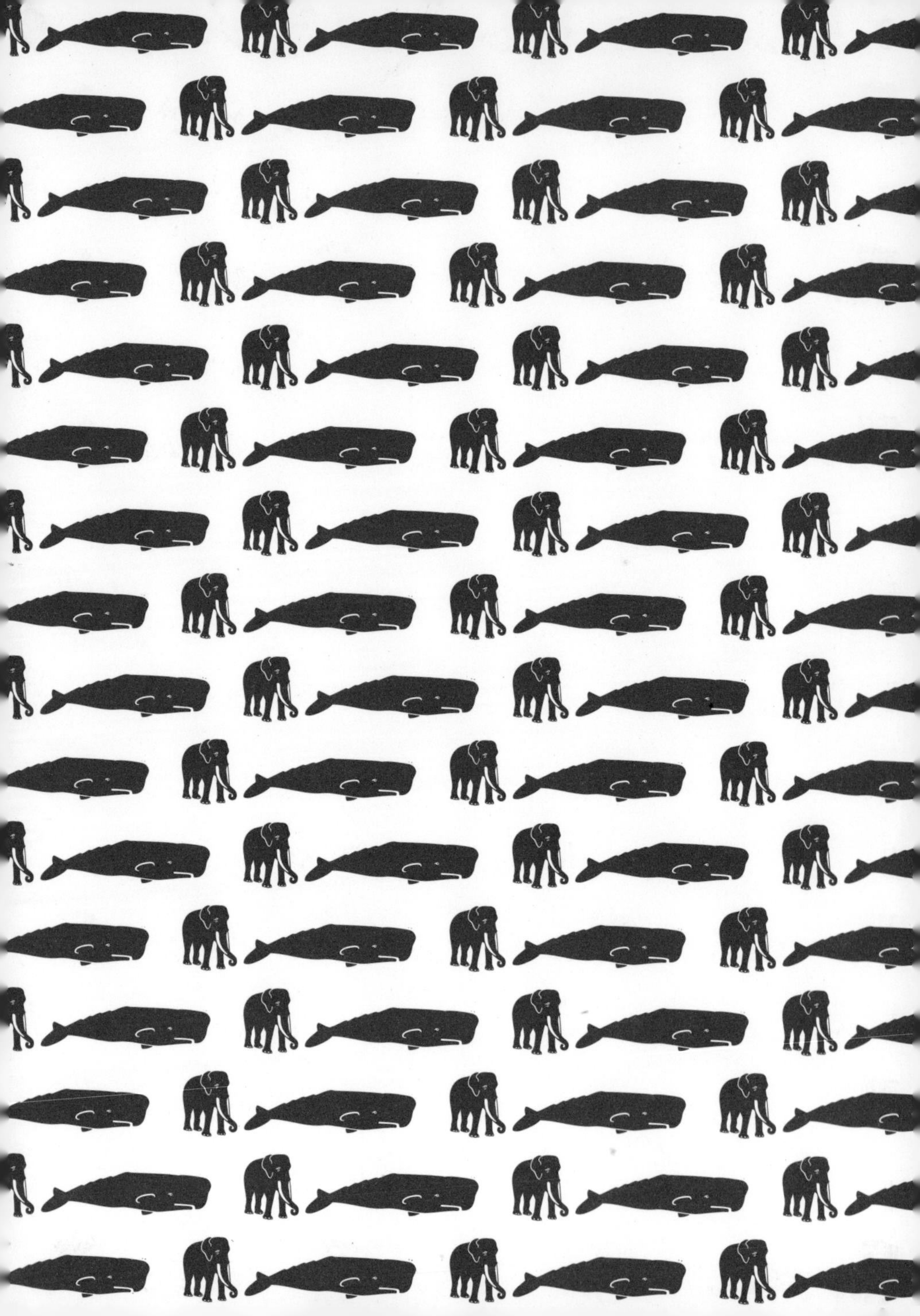